Water Issues in Himalayan South Asia

Amit Ranjan
Editor

Water Issues in Himalayan South Asia

Internal Challenges, Disputes and Transboundary Tensions

Editor
Amit Ranjan
Institute of South Asian Studies
National University of Singapore
Singapore, Singapore

ISBN 978-981-32-9613-8   ISBN 978-981-32-9614-5   (eBook)
https://doi.org/10.1007/978-981-32-9614-5

© The Editor(s) (if applicable) and The Author(s), under exclusive license to Springer Nature Singapore Pte Ltd. 2020
This work is subject to copyright. All rights are solely and exclusively licensed by the Publisher, whether the whole or part of the material is concerned, specifically the rights of translation, reprinting, reuse of illustrations, recitation, broadcasting, reproduction on microfilms or in any other physical way, and transmission or information storage and retrieval, electronic adaptation, computer software, or by similar or dissimilar methodology now known or hereafter developed.
The use of general descriptive names, registered names, trademarks, service marks, etc. in this publication does not imply, even in the absence of a specific statement, that such names are exempt from the relevant protective laws and regulations and therefore free for general use.
The publisher, the authors and the editors are safe to assume that the advice and information in this book are believed to be true and accurate at the date of publication. Neither the publisher nor the authors or the editors give a warranty, expressed or implied, with respect to the material contained herein or for any errors or omissions that may have been made. The publisher remains neutral with regard to jurisdictional claims in published maps and institutional affiliations.

Cover illustration: Marina Lohrbach_shutterstock.com
Cover design by eStudio Calamar

This Palgrave Macmillan imprint is published by the registered company Springer Nature Singapore Pte Ltd.
The registered company address is: 152 Beach Road, #21-01/04 Gateway East, Singapore 189721, Singapore

PREFACE

Defining a region is very difficult task. It is not static. Every scholar has own vague idea about it. Changing technology, geopolitical events, population movement and many other dynamics make one to define or re-define and construct or de-construct a region.[1] South Asia is one such region. Since the British colonial period, this region has been defined or re-defined to suit the political and academic interests of the individuals. For example, despite contiguity, Myanmar is not considered as a part of South Asia by majority of scholars and political commentators because it is not an active member of the South Asian Association for Regional Cooperation (SAARC), a regional organization that defines the political boundary of the region. However, a small number of scholars and commentators who give importance to history, contiguity and other related factors argue that Myanmar is very much part of South Asia. Likewise, many in Pakistan argues that the country is part of the West Asia instead of South Asia because of religious bonding. Then, there is a growing use of a term "extended neighbourhood" in India.[2] This entails

[1]Hagrety, D. T., & Hagerty, H. G. (2007). Reconstitution and Reconstruction of Afghanistan. In D. T. Hagrety (Ed.), *South Asia in World Politics* (pp. 113–133). Lanham, Boulder, New York, Toronto, and Oxford: Rowman and Littlefield.

[2]Shekhar, V. (2019). Rise of India's 'Extended Neighbourhood' Worldview. In A. Ranjan (Ed.), *India in South Asia: Challenges and Management* (pp. 235–252). Singapore: Springer.

vi PREFACE

comprehensive engagements with the countries falling within the ambit of "extended neighbourhood". The focus is also on extending the geographical scope of the region by terming it Southern Asia which includes countries such as Myanmar, etc.

Physically, as a region, Himalayan South Asia comprises north, east and northeast India, parts of Pakistan and Nepal and Bhutan. Afghanistan falls in Hindu Kush Himalayan which is considered as a part of Greater Himalayas. There is almost no presence of Himalaya in the deltaic Bangladesh, but the Chittagong Hill Tract is considered by some geographers as a range within the Greater Himalayan system.[3]

Despite such differences, all these countries depend on Himalayan and any action in its mountain ranges affects them. In South Asia, a large part of India, Pakistan, Afghanistan, Bangladesh and entire Nepal and Bhutan depends on the waters of Himalayan river system. Even parts of China and Myanmar depend on waters from Himalayan rivers. Some of the major river systems which flow in this region are Indus River System, Ganga River System, etc. As respective river systems intrinsically link the countries falling in their catchment areas, they can be an important factor to construct and define a region. In this work, although the editor has not made an attempt to theorize a region by taking water as an important factor, one of the objectives of this study is to at least sketch such definition.

In this book, instead of looking at Himalayan South Asia as a region which lies in the foothills of the Himalaya, water issues of the entire State have been dealt with because of four reasons: First, the River systems which originate in the Himalaya affect a large part of the State and not only the catchment areas. For example, Bangladesh is a deltaic country but the rivers such as Teesta, Ganga and others which originate in the Himalaya and are part of the Himalayan river system affect the larger part or almost the entire country. As these rivers are also transboundary in nature, they are cause for Bangladesh's water disputes with India.

Second, the water stress or scarcity or severe water scarcity in one part of the country has serious impact in other parts because of significance of water-related economic activities such as agriculture. For example, whenever production of grains are relatively low due to decline in water

[3]Banglapedia: National Encyclopaedia of Bangladesh. *Himalayas, The.* http://en.banglapedia.org/index.php?title=Himalayas,_The. Accessed June 21, 2019.

PREFACE   vii

availability in Punjab in Pakistan, it affects the entire country. Lower production negatively hits the agriculture-dependent economy and increases the prices of food grains. Third, disputes and tensions on sharing of waters from same river are mainly between the areas or units which lie in same floodplains and catchment areas. However, these catchment areas and floodplain are also part of the different sovereign States. This make the States instead of people living in those catchment areas and floodplain to make policies or take decisions on those waters. Once the state is engaged, it becomes an issue between the two sovereigns. For example, India–Pakistan tensions over sharing the waters from the Indus River System also aggravate water nationalism among people from the Central or South India who have nothing to do with waters from those rivers.

Fourth, water stress in one region affects the water behaviour of a country in different regions. For example, to deal with the water stress in its northern region, China is engaged in transfer of waters through South-to-North Water Diversion Project/South to North Water Transfer Project (SNWDP/SNWTP) or *Nanshuibeidiao*. The third line in this project called the Great Western Line is a matter of concern for India. If this line gets fully implemented, it is expected that waters from the Himalayan river system would also be transferred to different parts of China. Likewise, to deal with the water stress in different parts of the country, India is also engaged in interlinking some of its major rivers. This includes the Himalayan rivers. Keeping this reason in mind, a chapter dealing with water disputes in India and another one which discusses politics of regional water management in China have been included in this book.

Unfortunately, despite having a number of large river systems, Himalayan South Asia and China are not water secure. As shown in the *The United Nations World Water Development Report 2018*, some of the zones in South Asia may face severe water scarcity by 2050.[4] One of the reasons for the scarcity is rise in population which increases the gap between demand and supply. In Himalayan South Asia, India has a population of about about 1.35 billion, Pakistan's population as more than 216 million people, Bangladesh has about 163 million people, Nepal

---

[4] *The United Nations World Water Development Report 2018*. Nature-Based Solutions for Water. https://reliefweb.int/sites/reliefweb.int/files/resources/261424e.pdf. Accessed June 12, 2019, p. 12.

viii   PREFACE

has about 28.61 million people and Bhutan's population is 763,092. This is likely to increase further in coming years. Rising population will increase the demands for food, industrial goods and waters for domestic consumption. Like other parts of the world, in this region too, about 70% of total available waters are used for agriculture activities, 20% for industrial production and 10% for domestic consumption.[5] This percentage may vary a bit from one country to the other; however, the ratio of sector wise consumption out of total available waters in terms of quantity—maximum to minimum—remains the same.

Besides population, due to the phenomenon of climate change, glaciers—a source to most of the rivers of the region—are melting. This is causing floods even in non-rainy season. The phenomenon of climate change is also a reason for more precipitation during the rainy season which leads to severe floods. However, not all glaciers are melting. There are studies which claim that eastern and central Himalayas are retreating but western Himalayas, source of most Pakistani rivers, one of the severely stressed countries of the region, are stable and even increasing in size. [6]

In addition to declining availability and phenomenon of climate change, a large quantity of available waters are polluted, and therefore cannot be used for consumption. In the three most industrialized countries of the Himalayan South Asia region—Pakistan, Bangladesh and India—the untreated effluents from many industries are directly or after little treatment are released into the rivers or water bodies. Due to pollution, many Himalayan rivers or of their tributaries in India, Pakistan or Bangladesh are not suitable for any form of direct human touch.

There are also problems of management of water resources in the Himalayan South Asian countries. Most of them rely too much on the supply-side of water management. Big multipurpose hydro projects are constructed to manage the waters. Some of these projects cause displacement and are not ecologically friendly. These countries also lack technologies by which more crops can be produced from little drops of water. A large part of the region depends on water-intensive agriculture practice.

[5]Ibid., p. 11.

[6]The National Academy of Sciences, Himalayan Glaciers Climate Change, Water Resources, and Water Security, 2012. https://www.nap.edu/resource/13449/Himalayan-Glaciers-Report-Brief-Final.pdf. Accessed January 12, 2019.

To address some of their water concerns, the co-riparian States compete to have more quantity of waters. This causes tensions and disputes between them. Although there are treaties or Memorandum of Understanding signed between the riparian States, disputes remain or recur.

Taking the above-mentioned reasons in mind, an attempt has been made to look at the water issues in Afghanistan, India, Bangladesh, Bhutan, Nepal and Pakistan, and then analyse the transboundary waters tensions and disputes. The contributing authors have discussed water issues in the respective countries without talking in detail about the transboundary water disputes which has been discussed in Chapter "Domestic Water Stress, Transboundary Tensions and Disputes" of this book. It has to be noted that most of the co-riparian countries of the region studied in this book have water-related tensions and disputes which they have tried to manage through some form of water sharing arrangements between them. An example of it is Indus Waters Treaty between India and Pakistan. Despite having three wars (1965, 1971 and 1999) and series of political-cum-military tensions, the treaty is intact; however, a section of people and leaders from India now call for its revocation. Unlike India–Pakistan, India–China, India–Bangladesh, India–Nepal, India–Bhutan and Afghanistan–Pakistan have more of political tensions and differences rather than disputes, in technical sense, over their transboundary waters.

Writing acknowledgements are the most difficult job for any author. While working on this book the editor has been directly or indirectly helped by many people in different ways. It is difficult to acknowledge all of them in the given space. In particular, I have been helped by a few individuals whom I must acknowledge. I am thankful to Professor C. Raja Mohan, Mr. Hernaikh Singh, Dr. Dipinder Singh Randhawa, Dr. Rahul and Sylvia. All of them are my colleagues at the Institute of South Asian Studies, National University of Singapore.

I could not even imagine about this book without contributors. They have been very nice to me and stuck to the deadlines. I am grateful to all of them. Finally, I acknowledge Sandeep Kaur from Palgrave who gave me more time than I initially demanded.

Singapore, Singapore   Amit Ranjan

CONTENTS

Emerging Water Scarcity Issues and Challenges in Afghanistan   1
Fazlullah Akhtar and Usman Shah

Water Management in Bangladesh: Policy Interventions   29
Punam Pandey

Water Issues in Bhutan: Internal Disputes and External Tensions   51
Rajesh Kharat and Aanehi Mundra

The Politics and Policies of Regional Water Management in Southern China   77
Kris Hartley

Mapping the Water Disputes in India: Nature, Issues and Emerging Trends   103
Ruchi Shree

Multi-stakeholder Hydropower Disputes and Its Resolutions in Nepal   125
Sanju Koirala, Prakash Bhattarai and Sarita Barma

xii CONTENTS

Is Pakistan Running Dry? 153
Zofeen T. Ebrahim

**Domestic Water Stress, Transboundary Tensions
and Disputes** 183
Amit Ranjan

Bibliography 205

Index 207

NOTES ON CONTRIBUTORS

Dr.-Ing. Fazlullah Akhtar holds Ph.D. (Engineering) from the Center for Development Research (ZEF), University of Bonn (Germany) and is currently affiliated with the institute. Prior to ZEF, he was working with the UNFAO, USAID, German Agro Action and the Ministry of Higher Education in Afghanistan.

Sarita Barma is currently working as a Research Associate at Centre for Social Change (CSC) Nepal and mainly involved in labour migration and governance initiatives. Her research interests focus on natural resource management, psychological health and governance.

Dr. Prakash Bhattarai is currently acting as President of Centre for Social Change (CSC), a non-profit research and advocacy institute in Nepal. Dr. Bhattarai holds a Ph.D. in Peace and Conflict Studies from the University of Otago, New Zealand.

Zofeen T. Ebrahim is an independent journalist and currently the Pakistan editor for *The Third Pole* (www.thethirdpole.net) which is a multilingual platform dedicated to promoting information and discussion about the Himalayan watershed and the rivers that originate there. The project was launched as an initiative of chinadialogue (www.chinadialogue.net), in partnership with the Earth Journalism Network. It is a registered non-profit organisation based in New Delhi and London, with editors also based in Kathmandu, Beijing and Dhaka.

She has written extensively on development issues including climate change, water, energy, renewables, sanitation, health, women's health, diseases, etc., and how these impact our lives every day. She finds Pakistan a good place to be in as a journalist as there is absolutely no dearth of stories. "Every stone you turn, you will find a story under it," she says though few feel-good ones.

She contributes regularly to national English dailies like The Dawn (where she worked from 1994 to 2001) and *The News* as well as international media including the Inter Press Service, The Guardian, The Third Pole, University World News, Reuters, etc.

Dr. Kris Hartley is an Assistant Professor at the Department of Asian and Policy Studies, Education University of Hong Kong. He researches development policy in Asia with a focus on innovation and technology. Kris is also a Nonresident Fellow for Global Cities at the Chicago Council on Global Affairs and an Affiliated Scholar at the Center for Government Competitiveness at Seoul National University.

Prof. Rajesh Kharat is Dean of Humanities, University of Mumbai. Earlier, he was Professor at Centre for South Asian Studies, Jawaharlal Nehru University, New Delhi.

Dr. Sanju Koirala holds a Ph.D. in Human Geography from the University of Otago, New Zealand, with particular focus on hydropower induced displacement, and resettlement. Over the past 10 years, she has been involved in a research projects on issues around water resource, community forest, micro-enterprise, migration and youth policy.

Aanehi Mundra is a Ph.D. scholar at South Asian Studies, School of International Studies, Jawaharlal Nehru University, New Delhi.

Dr. Punam Pandey is a Ph.D. graduate from the University of Delhi (DU). Before joining the University of Free State, she was an Assistant Professor at DU. Punam is a Research Associate in the Institute for Reconciliation and Social Justice (IRSJ), University of the Free State, Bloemfontein, South Africa. She broadly works on issues of South Asia with a special focus on India and Bangladesh. She has written extensively on water politics from the perspective of International Relations.

Dr. Amit Ranjan is a Research Fellow at Institute of South Asian Studies, National University of Singapore, Singapore.

Mr. Usman Shah is a graduate of Monash University, Australia. He conducted his M.A. research on the local impacts of integrated water resource management interventions amongst communities in Kunduz and Takhar provinces of Afghanistan. He continues to contribute to research on development issues related to natural resources, conflict, migration and security.

Dr. Ruchi Shree teaches Political Science at Janki Devi Memorial College, University of Delhi, India. Her research interests include water as commons, development discourse, environmental movements, Gandhi politics and peace studies. She is also associated with several research and advocacy organizations viz. Environmental Law Research Society (ELRS), Jaladhikar Foundation, Parampara, Hanns-Seidel-Stftung (HSS-India), etc.

ABBREVIATIONS

ADB	Asian Development Bank
ANDMA	Afghanistan National Disaster Management Authority
BWDB	Bangladesh Water Development Board
CDA	*Critical Discourse Analysis*
CSO	Civil Society Organisations
CTGC	China Three Gorges Company
DoE	Department of Environment
FCDI	Flood Control Drainage Irrigation Projects
GDP	Gross Domestic Product
GLOF	Glacial Lake Outburst Flooding
GLOFs	Glacial Lake Outburst Floods
IAD	Institutional Analysis and Development
IMF	International Monetary Fund
INHURED	International Institute for Human Rights, Environment and Development
IRN	International River Network
ISWD	Inter-State Water Disputes Act
IWMI	International Water Management Institute
IWRM	Integrated Water Resources Management
MAIL	Ministry of Agriculture Irrigation and Livestock
MEW	Ministry of Energy and Water
MRRD	Ministry of Rural Rehabilitation and Development
NBA	Narmada Bachao Andolan
NEA	Nepal Electricity Authority
NEC	National Environment Commission
NECS	National Environment Commission Secretariat

xviii ABBREVIATIONS

NEP	New Economic Policy
NEPA	National Environmental Protection Agency
NITI Aayog	National Institution for Transforming India
NRPC	National River Protection Commission
NSP	National Solidarity Program
NWPo	National Water Policy (of Bangladesh)
NWRC	National Water Resources Council
PRD	Pearl River Delta
PSU	Public Sector Undertakings
RCP	Representative Concentration Pathways
RWL	Radius Water Limited
SASWE	Sustainability, Satellites, Water, and Environment
SES	Socio-Ecological Systems
SMEC	Snowy Mountains Engineering Corporation
SOPPECOM	Society for Promoting Participative Ecosystem Management
SSD	Sardar Sarovar Dam
TWM	Total Water Management
UN	United Nations
UNDP	United Nations Development Programme
UNEP	United Nations Environment Program
USD	United States Dollar
UTPCC	Upper Tamakoshi Peoples Concern Committee
VDC	Village Development Committee
WARPO	Water Resources Planning Organisation
WHO	World Health Organization
WIN	Water Intelligence Network
WSCS	West Seti Concern Society
WSHP	West Seti Hydropower Project
WUA	Water User Association

LIST OF FIGURES

Emerging Water Scarcity Issues and Challenges in Afghanistan

Fig. 1   Water availability and demand analysis in the Kabul river basin, Afghanistan (*Source* Akhtar, F. [2017], (PhD dissertation). University of Bonn. http://hss.ulb.uni-bonn. de/2017/4824/4824.pdf)   2

Fig. 2   Groundwater recharge (Uhl 2003; FAO 1996) and use (FAO 1996) in river basins of Afghanistan   16

Water Issues in Bhutan: Internal Disputes and External Tensions

Fig. 1   Riverside in Paro. Aanehi Mundra. April 2016   68

The Politics and Policies of Regional Water Management in Southern China

Fig. 1   Institutional analysis and development framework (Ostrom, E. [2007]. Institutional Rational Choice: An Assessment of the Institutional Analysis and Development Framework)   90

Fig. 2   Framework for transboundary water governance (Jensen, K., & Lange, R. [2013]. *Transboundary Water Governance in a Shifting Development Context* [p. 25, DIIS Report 2013:20]. Danish Institute for International Studies. https://www.diis. dk/en/research/transboundary-water-governance-in-a-shifting-development-context-0)   92

| Fig. 3 | Hybrid framework (Hartley, K. [2017]. Environmental Resilience and Intergovernmental Collaboration in the PRD. *International Journal of Water Resources Development, 34*[4], 525–546) | 93 |
| Fig. 4 | Investment in alternative supply (*Source* Author) | 94 |

Emerging Water Scarcity Issues and Challenges in Afghanistan

Fazlullah Akhtar and Usman Shah

INTRODUCTION

Afghanistan's location at the crossroads of several regions of Asia has placed it in a strategic role in geopolitical rivalries, which are increasingly playing out in natural resource contestations. The country's terrain is defined by the Iranian Plateau and the Hindu Kush mountains from which water drains through five major river basins. Except for the northern river basin, all of the four river basins are of transboundary nature and discharge water into the neighbouring countries: The Kabul River into Pakistan's Indus, the Hari Rod and the Helmand rivers to Iran, the Pyanj/Amu Darya into the former-Soviet Central Asian republics. The five major river basins of Afghanistan include the following[1] (Fig. 1).

1. The Kabul (Indus) river basin

[1] Kamal, G. M. (2004). *River Basins and Watersheds of Afghanistan*. Afghanistan Information Management Service (AIMS), 1–4.

F. Akhtar (✉)
Center for Development Research (ZEF), University of Bonn, Bonn, Germany
e-mail: fakhtar@uni-bonn.de

U. Shah
Monash University, Melbourne, Australia

© The Author(s) 2020
A. Ranjan (ed.), *Water Issues in Himalayan South Asia*,
https://doi.org/10.1007/978-981-32-9614-5_1

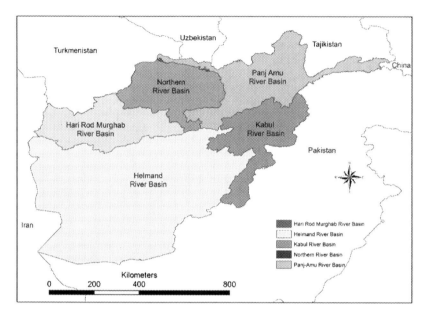

Fig. 1 Water availability and demand analysis in the Kabul river basin, Afghanistan (*Source* Akhtar, F. [2017], (PhD dissertation). University of Bonn. http://hss.ulb.uni-bonn.de/2017/4824/4824.pdf)

2. The Panj-Amu river basin
3. The Northern river basin
4. The Hari Rod-Murghab river basin
5. The Helmand river basin

Roughly 98% of surface water supplies in Afghanistan are used for agriculture, with the remaining used for domestic and industrial purposes.[2] Despite this high usage by the agricultural sector, the country remains greatly dependent upon food imports from neighbouring countries to fulfil local consumption. Scholarly research has suggested that this has been

---

[2]FAO. (2015). *Afghanistan: Geography, Climate and Population* (Food and Agriculture Organization of the United Nations). http://www.fao.org/nr/water/aquastat/countries_regions/afg/index.stm.

exacerbated by irrigation systems operating at below optimal standards.[3] Furthermore, the area of land in Afghanistan under irrigation/crop cultivation has not increased a great deal in the last decade.[4] An increasing population and the consequent increase in water demand for various purposes, along with climate change, pose threats to different sectors which require heightened attention by the authorities to respond to the growing conflicts and crises relevant to water resources. However, there are serious issues that stand in the way of effective water resource management. Institutional short-comings and overlapping jurisdictions and the overall 'weak state' context of Afghanistan have stood in the way of sound, research-based, management outcomes. Importantly, the internecine armed conflict that has characterized Afghanistan for four decades has prevented effective governance, development work, and provided the 'fog of war' to allow domestic actors and neighbouring countries to extract water unregulated.

Afghanistan has been designated as one of the countries suffering from the most intense physical water scarcity in the world[5]; it is projected to be suffering from extreme water scarcity by 2040.[6] Climate change scenarios for Afghanistan project worsening of the existing conditions, affecting crops, livestock, access potable water, or for industrial use, and likely also alter the hydrological regime of most of the watersheds across the country. Currently, other than potential impacts of climate change, ground and surface water pollution, poor water resources management and governance lead to inequality, inadequacy and unreliability during irrigation water distribution which results into inefficiency in irrigated agriculture and poor crop-water productivity.[7] Consequently, there are gaps between

[3] Akhtar, F., Awan, U., Tischbein, B., & Liaqat, U. (2018). Assessment of Irrigation Performance in Large River Basins Under Data Scarce Environment—A Case of Kabul River Basin, Afghanistan. *Remote Sensing, 10*(6), 972.

[4] Akhtar, F., Awan, U. K., Tischbein, B., & Liaqat, U. W. (2017). A Phenology Based Geo-Informatics Approach to Map Land Use and Land Cover (2003–2013) by Spatial Segregation of Large Heterogenic River Basins. *Applied Geography, 88*, 48–61.

[5] Comprehensive Assessment of Water Management in Agriculture. (2007). *Water for Food, Water for Life: A Comprehensive Assessment of Water Management in Agriculture*. London: Earthscan; Colombo: International Water Management Institute.

[6] World Resources Institute. https://www.wri.org/blog/2015/08/ranking-world-s-most-water-stressed-countries-2040. Accessed December 11, 2018.

[7] Akhtar, F., Awan, U. K., Tischbein, B., & Liaqat, U. (2018). Assessment of Irrigation Performance in Large River Basins Under Data Scarce Environment—A Case of Kabul River Basin, Afghanistan. *Remote Sensing, 10*(6), 972.

water supply and demand, which ultimately causes failure to meet the growing food demand across the country. Water losses occur during conveyance and application affecting agricultural water consumption. But it also affects future industrial and municipal developments, which are deemed to have enhanced water demands. There is thus, heightened scope for conflicts between water users at different reaches of a given watershed. Research has found that most irrigation systems in Afghanistan are experiencing acute water shortages and require attention in terms of management interventions.

Currently, the irrigation water distribution system among farmers located along the different reaches of canals in Afghanistan is supply based rather than crop-water-need based. As a result, there are structural inequalities in irrigation systems within each canals' network, as a result of which downstream farmers are not allocated longer irrigation times while the farmers located at the canal upstream divert more water and grow mostly cash crops which creates inequality and results into conflicts among farmers at different reaches. The presence of such inequalities in irrigation systems is supported by consistent reports of higher yields and cash returns for crops grown at the upstream, compared to those located at the downstream.[8] Further challenges and issues are elaborated more comprehensively in the following section.

CHALLENGES TO EFFECTIVE WATER RESOURCE GOVERNANCE

Climate Change

Although some research on climate change and its impacts on natural resources in Afghanistan has been conducted,[9] analysis has been limited and has not sufficiently captured the intersecting forces at play that will increasingly play out in the context of a changing climate.[10] Furthermore, previous

---

[8] Pain, A. *Land, Power and Conflict in Afghanistan: Seeking to Understand Complexity.* https://journals.openedition.org/remmm/7990. Accessed December 12, 2018.

[9] Ososkova, T., Gorelkin, N., & Chub, V. (2000, March). Water Resources of Central Asia and Adaptation Measures for Climate Change. *Environmental Monitoring and Assessment, 61*(1), 161–166.

[10] See Rasul, G., & Sharma, B. (2016). The Nexus Approach to Water-Energy-Food Security: An Option for Adaptation to Climate Change. *Climate Policy, 16*(6), 682–702.

research has found that the projections for climate change in Afghanistan must be bolstered considerably, as there is a historical dearth of meteorological data.[11] Recent studies have been able to point to a growing nexus of environmental and social forces being exacerbated by shifting climate trends. According to the United Nations Environment Program (UNEP), about 80% of conflicts in Afghanistan are related to resources like land and water and to food insecurity.[12] The ability of developing countries to mitigate the effects of climate change is an increasing area of study.[13]

Under the Representative Concentration Pathways (RCP 4.5), Afghanistan is projected to warm by approximately 1.5°C by 2050, followed by a period of stabilization and then additional warming of approximately 2.5°C until 2100.[14] However, under the more pessimistic scenario (RCP 8.5) the country will warm by approximately 3°C by 2050, with further warming by up to 7°C by 2100.[15] Under both scenarios higher temperature increases are expected at higher altitudes than the lowlands. In the Central Highlands and the Hindu Kush region, warming over the period from 2021–2050 is projected to range from 1.5 to 1.7°C compared to the base period (1976–2006), while in the lowlands the increase ranges from 1.1 to 1.4°C.[16] The increase of temperature in central highlands will negatively affect the snowfall pattern and amount which will ultimately challenge the amount of water available for irrigating the existing crops' cover.[8]

Such increases in temperature will certainly accelerate the rate of evapotranspiration across the country and most certainly this will negatively impact the hydrological cycle, affect the agricultural productivity, alter

---

[11] Savage, M., Dougherty, B., Hamza, M., Butterfield, R., & Bharwani, S. (2009). Socio-Economic Impacts of Climate Change in Afghanistan. Oxford, UK: Stockholm Environment Institute.

[12] Rasmussen, S. E. (2017). How Climate Change Is a 'Death Sentence' in Afghanistan's Highlands. *The Guardian.* https://www.theguardian.com/world/2017/aug/28/how-climate-change-is-death-sentence-afghanistan-highlands-global-warming. Accessed December 8, 2018.

[13] Mirza, M. M. Q. (2003). Climate Change and Extreme Weather Events: Can Developing Countries Adapt? *Climate Policy, 3*(3), 233–248.

[14] NEPA & UNEP. (2015). *Climate Change and Governance in Afghanistan* (p. 7). Kabul: National Environmental Protection Agency and United Nations Environment Programme.

[15] Ibid.

[16] Ibid.

the cropping pattern both in high and lowlands as well as challenge the availability of water resources. As per local perceptions, temperatures have risen over the past decades. Rain is already scarcer and more unpredictable. According to a report by the UNEP, the World Food Programme and the National Environmental Protection Agency (NEPA), the biggest climate hazards to Afghan livelihoods are drought caused by reduced rain, declining river flows due to reduced spring-time snowmelt in the highlands and floods caused by increased heavy spring rainfall, and riverine floods caused by heavier and faster upstream snowmelt in the highlands.

There is a projected decrease of precipitation during springtime (March–May) for the North, the Central Highlands and the East from 2006 to 2050 between 5 and 10%.[17] This decrease is offset by a slight increase of precipitation during autumn and wintertime (October–December) in these regions. There is also projected increase in precipitation during the winter season of approximately 10%.[18] For the arid South of the country, the models do not project significant trends for precipitation. In terms of changes to the frequency of annual rainfall, visual analysis of the scenarios does not reveal any significant change. Overall, the decrease of precipitation during springtime will have harmful consequences due to the fact that this is the period where plant growth takes place as a result of irrigation application. Decreased precipitation is projected in the East, North and Central Highlands—the regions of Afghanistan with the highest levels of agricultural productivity.[19] With increased temperature, the mountain snowmelt, which feed rivers throughout the basin, will be melting earlier which will decline the groundwater recharge and hence the major drinking water sources of the population will be adversely affected.[20] Shifting climatic patterns have therefore emerged as a major concern for fragile states such as Afghanistan, which suffer from overall 'weak state' context and will struggle to respond effectively as successive and cumulative factors coalesce.[21]

---

[17] NEPA & UNEP. (2015). *Climate Change and Governance in Afghanistan* (pp. 7–8). Kabul: National Environmental Protection Agency and United Nations Environment Programme.

[18] Ibid.

[19] Ibid.

[20] Ibid., 38.

[21] Mazo, J. (2009). Conflict, Instability and State Failure: The Climate Factor. *The Adelphi Papers, 49*(409), 87–118.

Population Growth and Increased Water Demand

High rates of natural population growth,[22] as well as increased returns of former refugees especially from Pakistan and Iran,[23] have exacerbated strain on already stressed resource supplies.[24] Analysts have long understood that a more peaceful Afghanistan would lead to increased claims on upstream water resources at the expense of downstream users.[25] However such concerns have been relegated by the need to fulfil potable water consumption needs. Kabul is one of the fastest growing cities in the world.[26] Recent estimates put the population of the city at 3.9 million people,[27] but it is projected to host a population of nine million by 2050.[28] Although Kabul residents' use of water is reportedly less than that in most other Asian cities at 40 liters per day, this demand is projected to grow along with the projected rise in population.[29]

However, increased demand for scarce water resources has not been the result solely of domestic consumption needs. Excessive commercial and domestic pumping has caused a sharp decline in the groundwater level

[22] Central Statistics Organizations. *Analysis of Population Projections 2017–18.* http://cso. gov.af/en/page/demography-and-socile-statistics/demograph-statistics/3897111.

[23] Returning Home, Afghans Continue to Face Challenges in Rebuilding Their Lives—UN Agencies. *UN News,* April 12, 2018. https://news.un.org/en/story/2018/04/1007131; Bjelnica, J., & Ruttig, T. (2017, May 19). Voluntary and Forced Returns to Afghanistan in 2016/17: Trends, Statistics and Experiences. *Afghanistan Analysts Network.* https://www.afghanistan-analysts.org/voluntary-and-forced-returns-to-afghanistan-in-201617-trends-statistics-and-experiences/.

[24] Habib, H. (2014). Water Related Problems in Afghanistan. *International Journal of Educational Studies,* 1(3), 137–144.

[25] Rubin, B. R., & Armstrong, A. (2003). Regional Issues in the Reconstruction of Afghanistan. *World Policy Journal,* 20(1), 31–40; Glantz, M. H. (2005). Water, Climate, and Development Issues in the Amu Darya Basin. *Mitigation and Adaptation Strategies for Global Change,* 10(1), 23–50; Kuzmits, B., (2006). *Cross-Bordering Water Management in Central Asia* (ZEF Working Paper Series No. 66).

[26] City Mayors Statistics. *The World's Fastest Growing Cities and Urban Areas from 2006 to 2020.* http://www.citymayors.com/statistics/urban_growth1.html.

[27] *Estimated Population of Kabul City by District and Sex 2017–18.* Kabul: Central Statistics Office, 2018. http://cso.gov.af/en/page/demography-and-socile-statistics/demograph-statistics/3897111.

[28] Haidary, M. S. (2018, March 28). As Kabul Grows, Clean Water a Step Toward State Legitimacy in Afghanistan. *In Asia: Weekly Insights and Analysis.* https://asiafoundation.org/2018/03/28/kabul-grows-clean-water-step-toward-state-legitimacy-afghanistan/.

[29] Ibid.

8 F. AKHTAR AND U. SHAH

in recent years. According to the USGS, the groundwater table in Kabul declined by an average of 1.5 meters per year during 2004–2012.[30] Not only are groundwater aquifers dropping but the quality of water being extracted is also decreasing.[31] Lack of effective regulation and excess pumping has caused thousands of wells to go dry, requiring deepening or replacement.[32]

Contamination of underground water from domestic, agricultural and industrial wastewater released into the Kabul River also poses a grave health concern, with the majority of the shared water points and wells in the capital left contaminated today. So improved access to water has not necessarily translated into better access to *safe* water.[33]

It has long been understood by the corpus of research that demand for water resources in Afghanistan will rise if and when conflict situation stabilizes, allowing for development activities to ramp up as population growth rises.[34] Although Afghanistan has not seen the expected outcome for widespread peace and an end to conflict since large-scale international intervention in 2001, the population rise and development activities have pushed ahead. Conflict has therefore added a layer of exacerbation to issues with sustainability of use. In fact, research has repeatedly concluded that many local conflicts across the country are fought over water resources,[35] and the root cause was often the breakdown of customary water governance practices driven by increased population pressure on ever-scarce

[30] Mack, T. J. (2018). Groundwater Availability in the Kabul Basin, Afghanistan. In A. Mukherjee (Ed.), *Groundwater of South Asia* (pp. 23–35). Springer Hydrogeology. Singapore: Springer.

[31] Saryabi, A. H., Noori, A. R., Wegerich, K., & Kløve, B. (2017). Assessment of Water Quality and Quantity Trends in Kabul Aquifers with an Outline for Future Drinking Water Supplies. *Central Asian Journal of Water Research, 3*(2), 3–11.

[32] Hasib, S. (2017, February 28). Afghan Capital's Thirsty Residents Dig Deep to Combat Drought, Oversuse. *Reuters.* https://www.reuters.com/article/us-afghanistan-water/afghan-capitals-thirsty-residents-dig-deep-to-combat-drought-overuse-idUSKBN1670FO.

[33] See Thomas, V. (2015). *Household Water Insecurity: Changing Paradigm for Better Framing the Realities of Sustainable Access to Drinking Water in Afghanistan.* Kabul: Afghanistan Research & Evaluation Unit. https://areu.org.af/wp-content/uploads/2016/02/1522E-Household-Water-Insecurity.pdf. Accessed November 5, 2018.

[34] Sievers, Eric W. (2001). Water, Conflict, and Regional Security in Central Asia. *New York University Environmental Law Journal, 10,* 356.

[35] Pain, A. (2013, June). Land, Power and Conflict in Afghanistan: Seeking to Understanding Complexity. *Revue des Mondes Musulmans et de la Méditerranée, 133,* 63–81; Thomas, V., Azizi, M. A., & Ghafoori, I. (2013). *Water Rights and Conflict Resolution Processes in*

EMERGING WATER SCARCITY ISSUES AND CHALLENGES IN AFGHANISTAN   9

water resources.[36] Village-based case studies of rural Afghan dynamics have also found that if overall prosperity levels among the growing population can be raised even marginally, then conflicts over irrigation water distribution can be mitigated, even if issues with inequity in water access are not significantly changed.[37]

Poor Performance of the Irrigation Water Management
The United States' Department of Homeland Security study (2012) highlighted some of the key issues and projected problems with major river basins in the world that are strategically important to the United States interests, often as a result of their transboundary status. Among the river basins highlighted for their potential experiencing serious issues, includes the Amu Darya (which includes the Panj-Amu and Northern River basins of Afghanistan) and Indus (which includes the Kabul River Basin on Afghan territory). Before 2040, the Indus Basin (i.e. Kabul river basin in Afghanistan), is expected to experience chronic food insecurity and reduced resilience to flooding and drought as a result of poor water management, inefficient agricultural practices, soil salinization and greater variability in water availability.[38] Whereas, the Amu Darya River basin is considered likely to experience degradation in food security, increased regional tensions over water sharing before 2040 as a result of inadequate water sharing agreements, poor water quality and disruption of flows as well as poor water management practices.[39] At present, overall irrigation efficiency in

---

Afghanistan: The Case of the Sar-i-Pul Sub-basin. Kabul: Afghanistan Research and Evaluation Unit; Thomas, V. (2014). *Unpacking the Complexities of Water Conflicts Resolution Processes in Afghanistan*. Kabul: Afghanistan Research and Evaluation Unit.

[36] Pain.

[37] Ibid.

[38] Intelligence Community. (2012). *Global Water Security: Intelligence Community Assessment*. Washington, DC: Department of Homeland Security. https://www.dni.gov/files/documents/Newsroom/Press%20Releases/ICA_Global%20Water%20Security.pdf.

[39] Intelligence Community. (2012, February 2). *Global Water Security: Intelligence Community Assessment*, (ICA 2012-08), DIA, NSA, CIA, NGIA, BIR, FBI, OIA/Department of Homeland Security, ODNI/National Counterterrorism Center. https://www.dni.gov/files/documents/Newsroom/Press%20Releases/ICA_Global%20Water%20Security.pdf. Accessed December 24, 2018.

Afghanistan is in the range of 25–30% for both modern and traditional irrigation systems.[40] These losses are caused by earthen conveyance systems, operational losses, unlevelled lands, etc. Along the downstream portions of the Kabul River Basin, the mean application efficiency at a well-organized irrigation system is in the range of 46%,[41] variations do occur dependent on the availability of water at different crop growth stages.

As with other river basins in the region, the Kabul River Basin faces myriad and intersecting issues of governance, management, and development leading to inequity, inadequacy and unreliability of irrigation water distribution. Such outcomes are especially the case during the peak demand period that result in inefficient irrigated agriculture and poor crop-water productivity. The inefficiency in the irrigation system leads to larger voids between water demand and availability which retards the sufficiency in food production across the country.[42] By consuming around 98% of the surface water supplies for irrigated agriculture,[43] the country's dependency on food imports is illustrative of the insufficiency and inefficiency of the irrigation system, as well as the responsible institutions, in fulfilling local food demand.[44] The consumption of the highest amount of surface water supplies for agriculture purpose under high conveyance and on-farm losses also hampers the development of other water using sectors that are expected to be having increased water requirements in case of industrial and municipal developments in the near future.

The irrigation practices in all the river basins are conventional with minimum modifications; according to a recent study[45] carried out in the Kabul

---

[40] Qureshi, A. S. (2002). *Water Resources Management in Afghanistan: The Issues and Options* (Working Paper No. 49). Colombo: International Water Management Institute.

[41] Jalil, A., & Akhtar, F. (2018). *Performance Evaluation of the Irrigation System in Lower KabulRiver Basin, Afghanistan.* Tropentag.

[42] Akhtar, F., Awan, U. K., Tischbein, B., & Liaqat, U. W. (2018). Assessment of Irrigation Performance in Large River Basins Under Data Scarce Environment—A Case of Kabul River Basin, Afghanistan. *Remote Sensing, 10*(60), 972.

[43] FAO. *Afghanistan.* http://www.fao.org/nr/water/aquastat/countries_regions/AFG/AFG-CP_eng.pdf. Accessed December 24.

[44] Akhtar, F., Awan, U. K., Tischbein, B., & Liaqat, U. W. (2018). Assessment of Irrigation Performance in Large River Basins Under Data Scarce Environment—A Case of Kabul River Basin, Afghanistan. *Remote Sensing, 10* (60), 972.

[45] Ibid.

River Basin, all of its sub-basins are experiencing water stress. This therefore needs to be addressed and managed in order to meet the current and projected water and food demand in the country.

Intersectoral Competition

Estimations from 1998 placed withdrawal for agriculture purpose from Afghanistan's surface water supplies at roughly 98%, and industrial and municipal consumption or domestic use at 1% each.[46] With increase in population, industrial water demand is also increasing[47] which is required for independency, economic stability and revenue development which highlights the need for more water. A recent study on domestic water consumption in Kandahar city found that with a rising urban population, few households had access to secure sources of water and the government was unable to meet existing consumption needs through official water supplies. This ultimately, places pressure on existing agricultural water use.[48] Another emerging issue is the increasing understanding that hydropower, including micro hydropower plans placed along watercourses could offer multifaceted solutions to many of Afghanistan's development challenges.[49]

Increased domestic demand has the potential to create future transboundary conflict as less water remains in shared watercourses. Some research has pointed to the need for increased cooperative management of transboundary water resources in the region.[50] For example, in order

---

[46]FAO. (2018). *Afghanistan: Geography, Climate and Population* (Food and Agriculture Organization of the United Nations [FAO]). http://www.fao.org/nr/water/aquastat/countries_regions/afg/index.stm.

[47]Aini, A. (2007). Water Conservation in Afghanistan. *Journal of Developments in Sustainable Agriculture, 2*, 51–58).

[48]Haziq, M. A., & Panezai, S. (2017). An Empirical Analysis of Domestic Water Sources Consumption and Associated Factors in Kandahar City, Afghanistan. *Resources and Environment, 7*(2), 49–61.

[49]Rostami, R., Khoshnava, S. M., Lamit, H., Streimikiene, D., & Mardani, A. (2017). An Overview of Afghanistan's Trends Toward Renewable and Sustainable Energies. *Renewable and Sustainable Energy Reviews, 76*, 1440–1464.

[50]McKinney, D. C. (2004). Cooperative Management of Transboundary Water Resources in Central Asia. In *In the Tracks of Tamerlane: Central Asia's Path to the 21st Century*. Washington, DC: National Defense University, 187–220.

to address water scarcity for Kabul, the city's master plan calls for the construction of the Shahtoot Dam,[51] which is projected to provide 146 million cubic meters of potable water for two million residents of Kabul and enough water to irrigate 4000 hectares of land.[52] However, the Shahtoot Dam would cause up to a 17% drop in flows in the lower Kabul River on which neighbouring Pakistan's Peshawar Valley relies for agricultural production.[53]

Competition for water between rural and urban communities is also increasing, leading to heightened tensions and increased potential for conflict. The rural communities, besides rather limited domestic consumption, usually require water for irrigation purpose and livestock while the urban set up dominantly need it largely for domestic and industrial consumption.

Irrigation Water Distribution and Upstream-Downstream Farmers' Interaction

Afghanistan has largely maintained a traditional, community-based water resource management system for irrigation purposes. The canal water distribution is traditionally managed through a *mirab* (water master). The *mirab* distributes water according to the land area cultivated by individual farmers, irrespective of the water demanded by their chosen crop.[54] The entire irrigation system is supply based; water users either do not consider

---

[51] Amin, M., & Adeh, E. H. (2017, August 22). Water Crisis in Kabul Could be Severe If not Addressed. *The SAIS Review of International Affairs.* http://www.saisreview.org/2017/08/22/water-crisis-in-kabul-could-be-severe-if-not-addressed/; Hessami, E. (2014, November 13). Afghanistan's Rivers Could Be India's Next Weapon Against Pakistan. *Foreign Policy.* https://foreignpolicy.com/2018/11/13/afghanistans-rivers-could-be-indias-next-weapon-against-pakistan-water-wars-hydropower-hydrodiplomacy/.

[52] Hessami, E. (2014, November 13). Afghanistan's Rivers Could Be India's Next Weapon Against Pakistan. *Foreign Policy.* https://foreignpolicy.com/2018/11/13/afghanistans-rivers-could-be-indias-next-weapon-against-pakistan-water-wars-hydropower-hydrodiplomacy/.

[53] Ibid.

[54] Abdullayev, I., Mielke, K., Mollinga, P., Monsees, J., Schetter, C., Shah, U., & Ter Steege, B. (2009). Water, War and Reconstruction Irrigation Management in the Kunduz Region, Afghanistan. In M. Arsel & M. Spoor (Eds.), *Water, Environmental Security and Sustainable Rural Development: Conflict and Cooperation in Central Eurasia.* New York: Routledge.

the crop water requirements or are otherwise not aware of it.[55] Consequently, fields of standing crops are routinely flooded by upstream users, regardless of whether this is more than the optimal requirement or of the resulting water scarcity for farmers downstream. Meanwhile there is inadequate intervention by the relevant line ministries to implement water-sharing between different users as per the crop water demand while legally there government or line ministry can't bind the farmers to grow a specific crop. As a result, the appointment of *mirabs* as arbiters of water distribution decisions can become highly politicized and a potential flashpoint for conflict for communities sharing water resources.[56] Such local politics are made more problematic in ethno-linguistically diverse areas, such as Kunduz or Takhar provinces, where ethnic or political rivalries have reportedly been causal in contestations over irrigation water distribution decisions.[57]

In times of plentiful water flow in a canal, usually all of the farmers extracting water from it will receive their share of water more or less proportional to the area of land over which they have tenure.[58] However, in cases of water scarcity, the upstream farmers are able to extract the amount of water required for the irrigation of their crops, while the downstream farmers remain deprived of water to irrigate their lands. This causes reductions in productivity, crop failure, farmer debt and often leads to disputes with upstream communities.[59] Aggravating these conditions is that there is no such rule in the country for farmers to cultivate alternative crops with low water requirements in case of water scarcity, unless decided so by the farmers themselves. This leads to typical behavior among conventional

---

[55]Ter Steege, B. (2008). *Infrastructure and Water Distribution in the Asqalan and Sufi-Qaryateem Canal Irrigation Systems in the Kunduz River Basin* (ZEF Working Paper Series, No. 69). https://www.econstor.eu/bitstream/10419/88342/1/772774994.pdf.

[56]Mielke, K., Abdullayev, I., & Shah, U. (2010). The Illusion of Establishing Control by Legal Definition: Water Rights, Principles and Power in Canal Irrigation Systems of the Kundz River Basin, Afghanistan. In I. Eguavoen & W. Laube (Eds.), *Negotiating Local Governance; Natural Resource Management at the Interface of Communities and the State* (pp. 181–210). Munster: LIT Verlag.

[57]Abdullaev, I., & Shah, U. (2011). Community Water Management in Northern Afghanistan: Social Fabric and Management Performance. *International Journal of Environmental Studies, 68*(3), 333–334.

[58]Thomas, V., Mumtaz, W., & Azizi, M. A. (2012). *Mind the Gap? Local Practices and Institutional Reforms for Water Allocation in Afghanistan's Panj-Amu River Basin.* Afghanistan Research and Evaluation Unit, European Commission.

[59]Abdullayev and Shah.

farmers whereby they extract as much water as they possibly can, regardless of the water requirements of their crop. This is one of the key factors driving tensions and disputes between upstream and downstream water users.

The upstream water users usually tend to cultivate cash crops which requires them to divert more water for the entire growth period. Interference by power players also creates imbalances in the relations between upstream and downstream users.[60] In multipurpose canals, the situation could be even worse among different users of water in various reaches of the canal. Furthermore, there is widespread disaffection among different water users from ministry officials due to the lack of tangible improvement in other issues that hamper equitable and sustainable water distribution, including resolving critical issues of flood and emergency repairs as well as infrastructure rehabilitation.[61]

Unsustainable Groundwater Withdrawal and Reduced Natural Recharge

Groundwater is the key resource for domestic water consumption in Afghanistan and its project future availability has raised concerns due to population growth in recent years and the potential impacts of climate change on water resources. Research conducted in the aftermath of the international military intervention in 2001, suggested that groundwater pumping could take advantage of inexpensive deep drilling technology and deep well pumps for the rapid installation of wells to extract water for irrigation purposes.[62] Widespread groundwater extraction has caused the water table drop across the country.[63] According to recent studies, the groundwater level in Kabul city declines up to 1.5 meters per year which underlines the concern for the sustainability of groundwater resources amid growing demands for water.[64] The Afghanistan National Disaster Management Authority (ANDMA) has warned that groundwater reserves in Kabul

---

[60] Ter Steege.

[61] Ibid.

[62] Uhl and Tahiri, 34.

[63] Lashkaripour, G. R., & Hussaini, S. A. (2008). Water Resource Management in Kabul River Basin, Eastern Afghanistan. *The Environmentalist, 28*(3), 253–260.

[64] Mack, T. J. (2018). Groundwater Availability in the Kabul Basin, Afghanistan. In *Groundwater of South Asia* (pp. 23–35). Singapore: Springer.

will run dry within the coming decade as a result of overexploitation.[65] The unsustainable exploitation of groundwater resources is made more problematic as it also contributes to surface water and air pollution.

There is considerable illegal groundwater withdrawal in all of the river basins of the country. This occurs for diverse commercial purposes such as for car cleaning, industrial inputs, as well as for municipal needs.[66] While relevant public institutions, including the office of the Minister of Water and Energy, have publicly highlighted such illegal extraction as a concern, they are unable to halt the practice. Prosecution of such cases is hindered as much by an anaemic legal environment as by a lack of technical measurements of water extraction.

Throughout the country's five major river basins, the estimated recharge of groundwater is higher compared to the multipurpose usage. The highest rate of groundwater withdrawal is at Helmand and Kabul river basin while the highest recharge occurs at the Amu Darya river basin[67] (see Fig. 2). Groundwater extraction in the lower Helmand basin has ramped up to such a degree that between 2017 and 2018, an estimated 29,000 hectares of desert land has been reclaimed for agricultural purposes. This is attributed to increased access to technology, low land prices and has gone on despite the ongoing insecurity in the area.

The projected climate change scenarios foresee a higher evapotranspiration rate and reduced precipitation. This will likely further impede groundwater recharge which will, in turn, affect groundwater availability and ultimately the drinking water supply system thereby directly impacting the lives of the local population.

Dismal State of Water Resource Infrastructure
Afghanistan is an upstream country with no legal obligation to deliver water to the downstream countries, except Iran, with whom a treaty was signed in 1973. Despite this status, Afghanistan has not been able to fulfil the primary

---

[65] Amiri, S. (2018, July 13). Kabul's Underground Water Reserves 'To Dry Up Within Year. *Tolo News.* http://prod.tolonews.com/afghanistan/kabul%E2%80%99s-underground-water-dry-within-years-official.

[66] Felbab-Brown, V. (2017, March) "Water Theft and Water Smuggling" Growing Problem or Tempest in a Teapot?" *Foreign Policy at Brookings.* https://www.brookings.edu/wp-content/uploads/2017/03/fp_201703_water_theft_smuggling.pdf.

[67] Uhl & Tahiri, 34; FAO. (1996, November). Afghanistan, Promotion of Agricultural Rehabilitation and Development Programmes, Water Resources and Irrigation (FAO Project TCP/AFG/4552).

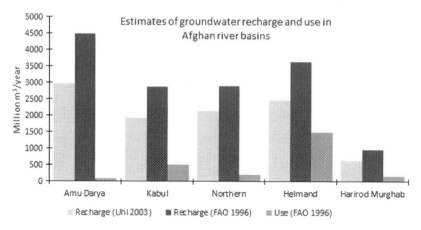

Fig. 2 Groundwater recharge (Uhl 2003; FAO 1996) and use (FAO 1996) in river basins of Afghanistan

water needs of its population. The Ministry of Energy and Water (MEW) has placed blame on the country's water woes for its vulnerability to climate change, which has manifested in a 62% drop in rainwater.[68] However, it has also been acknowledged that insufficient infrastructure has hindered Afghanistan's capacity to mitigate the negative consequences of climate change, for instance by storing snowmelt in reservoirs.[69] Therefore, the trend towards reduced precipitation and recharge has been compounded by a chronic lack of infrastructure and has become the primary hindrance to improving access to water for the population of Afghanistan.[70] As a consequence, of inadequate infrastructure in the Indus River Basin, there has, in turn, been a degradation of regional food security and reduced resilience to floods and droughts.

---

[68] Parwani, S. (2018, October 10). Is Water Scarcity a Bigger Threat Than the Taliban in Afghanistan? *The Diplomat.* https://thediplomat.com/2018/10/is-water-scarcity-a-bigger-threat-than-the-taliban-in-afghanistan/.

[69] Ibid.

[70] Peter, A. (2010). Afghanistan's Woeful Water Management Delights Neighbors. *Christian Science Monitor.* Retrieved from http://www.csmonitor.com/World/Asia-South-Central/2010/0615/Afghanistan-s-woeful-water-management-delights-neighbors.

The irrigated area in Afghanistan is managed under both traditional and modern systems. Protracted conflict and lack of investment in the irrigation sector caused degradation of the irrigation schemes during the past few decades.[71] Traditional infrastructure, especially in southern and western Afghanistan has relied on underground well and canal networks, known as *karez*. These systems would channel water for agriculture and domestic uses in even the driest corners of Afghanistan. They would also reduce evaporative losses. Such traditional systems have deteriorated during decades of protracted conflict.[72] Improving the situation of degraded or non-existent water infrastructure has been central to international development interventions since 2001.[73] Much of these have focused on improving access to safe drinking water.[74] Since the early 2000s, hundreds of millions of US dollars has been disbursed by the international community through grants of the National Solidarity Program (NSP) aimed at addressing small-scale irrigation infrastructure needs.[75]

At the same time, easier and more affordable access to technology such as drilling machinery and submersible pumps has led to the unregulated expansion of deep-well infrastructure and extraction from groundwater resources.[76] As a result of poor infrastructural and technical capacities, the overall irrigation efficiency of the system ranges between 25 and 30% for both modern and traditional systems.[77] The losses in the irrigation

---

[71] Rout, B. (2008). *Water Management, Livestock and the Opium Economy: How the Water Flows: A Typology of Irrigation Systems in Afghanistan.* Kabul: Afghanistan Research and Evaluation Unit.

[72] Hussain, I., Abu-Rizaiza, O. S., Habib, M. A. A., & Ashfaq, M. (2008). Revitalizing a Traditional Dryland Water Supply System: The Karezes in Afghanistan, Iran, Pakistan and the Kingdom of Saudi Arabia. *Water International, 33*(3), 333–349.

[73] MIWRE. (2004). *A Strategic Policy Framework for the Water Sector.* Kabul: MIWRE.

[74] Koch, D. (2009). *Improving Access to Safe Water, Sanitation & Hygiene for Children in Afghanistan.* Kabul: UNICEF. https://www.unicef.org/wash/afghanistan_48619. html; UN Water. (2015). *Afghanistan: Sanitation, Drinking-water and Hygiene Status Overview.* Geneva: UN Water. https://www.who.int/water_sanitation_health/monitoring/investments/afghanistan-7-1-16.pdf.

[75] Reich, D., & Pearson, C. (2013). Irrigation Outreach in Afghanistan: Exposure to Afghan Water Security Challenges. *Journal of Contemporary Water Research & Education, 149*(1), 33–40.

[76] see Mansfield.

[77] Qureshi, A. S. (2002). *Water Resources Management in Afghanistan: The Issues and Options* (Working Paper No. 49). Colombo: International Water Management Institute.

system (both modern and conventional) are driven by high on-farm distribution losses (i.e. over irrigation and poorly levelled farm units), high conveyance losses due to earthen channels as well as operational losses in modern schemes with lined conveyance canals.[78]

Large-scale dam infrastructure is regarded as problematic for myriad reasons. Firstly, the capacity of large water storage installations is limited and the efficiency of the few large reservoirs that exist has been hindered by heavy siltation due to heavy deforestation in watersheds. Furthermore, a risk averse investment climate has bred hesitation among of donor agencies in assisting Afghanistan with building large-scale water infrastructure, such as dams, reservoirs and big canals. This is largely rooted in the transboundary nature of most of the Afghan watersheds.[79] Diversion possibilities do exist in key transboundary basins e.g. Kabul/Indus Basin in Kunar province (shared with Pakistan), Amu Darya (shared with former-Soviet Central Asian states) and Helmand river basin (shared with Iran). However, the target areas in which key infrastructure need to be engineered and constructed are, for the most part, outside of the central government's control. Therefore, there is a security risk to government employees attempting to conduct background surveying, let alone commence key infrastructure construction works. Although development rhetoric has held that securing the country's access to water was tied to its overall security due to ongoing insecurity, institutional weakness hindering management and planning capacity, lack of transboundary frameworks for sharing, and issues with finances.[80] Attempts by Afghanistan to build dam infrastructure on the Helmand and Hari Rod and Kabul rivers have led to tensions with downstream Iran and Pakistan.[81] Research has conducted on the hydrological effects of potential future dam infrastructure projects in Afghanistan on downstream users in neighbouring countries and have led to calls for

[78] Ibid.

[79] Mashal, M. (2012, December 3). What Iran and Pakistan Want from the Afghans: Water. *Time*. https://pulitzercenter.org/reporting/what-iran-and-pakistan-want-afghans-water.

[80] Ahlers, R., Brandimarte, L., Kleemans, I., & Sadat, S. H. (2014). Ambitious Development on Fragile Foundations: Criticalities of Current Large Dam Construction in Afghanistan. *Geoforum, 54*, 49–58.

[81] Ramachandran, S. (2018, August 20). India's Controversial Afghanistan Dams. *The Diplomat*. https://thediplomat.com/2018/08/indias-controversial-afghanistan-dams/; Aman, F. (2013, January 7). Afghan Water Infrastructure Threatens Iran, Regional Stability. *Al-Monitor*. https://www.al-monitor.com/pulse/originals/2013/01/afghanwatershortageiranpakistan.html.

governments in these countries to take advantage of delays with pushing ahead with plans for water storage infrastructure in Afghanistan, caused by insecurity and lack of donor funds, by building their own reservoirs to mitigate against future water losses.[82] Furthermore, there have been widely reported claims that neighbouring countries have fomented unrest, including aiding armed opposition actors such as the Taliban, to attack and deter construction of large-scale, upstream infrastructure.[83] This climate has led to renewed calls for formal water-sharing agreements, brokered and supported by donor institutions such as the World Bank, with Afghanistan's neighbours to allow for cooperation around the construction of large-scale water infrastructure.[84] Iran has been accused of using militants to disrupt the Indian-funded Salma Dam, and Pakistan is accused of doing the same to render infrastructure projects in the Kabul Basin, such as the Indian-funded Shahtoot Dam, untenable.[85]

This nexus of insecurity, water scarcity and contestation and underdevelopment is shifting somewhat. Aside from more apparent willingness on the part of donor agencies to consider funding large-scale infrastructure,[86] peace talks with the Taliban have involved neighbouring countries, which will in turn affect their alleged role in obstructing upstream infrastructure construction. At the same time, the Afghan government has recognized that perhaps small-scale infrastructure, such as micro-hydropower

[82]Pervaz, I., & Khan, S. *Brewing Conflict over Kabul; Policy Options for Legal Framework*. ISSRA Papers 2014. https://ndu.edu.pk/issra/issra_pub/articles/issra-paper/ISSRA_Papers_Vol6_IssueII_2014/03-Brewing-Conflict-over-Kabul-River.pdf.

[83]Suleiman. (2018, March 27). Tehran Uses Taliban to Target Afghan Infrastructure Projects, Weaken Kabu. *Salaam Times.* http://afghanistan.asia-news.com/en_GB/articles/cnmi_st/features/2018/03/27/feature-01.; Mudabber, Z. (2016, November 11). Afghanistan's Water-Sharing Puzzle. *The Diplomat.* https://thediplomat.com/2016/11/afghanistans-water-sharing puzzle/.

[84]Mudabber (2016).

[85]Jain, R. (2018, July 17). In Parched Afghanistan, Drought Sharpens Water Dispute with Iran. *Reuters.* https://www.reuters.com/article/us-afghanistan-iran-water/in-parched-afghanistan-drought-sharpens-water-dispute-with-iran-idUSKBN1K702H.

[86]The World Bank. (2018, May 14). *Afghanistan Resurrects Its Largest Hydropower Plant Toward a Brighter Future.* https://www.worldbank.org/en/news/feature/2018/05/14/largest-plant-restarts-operations-in-first-step-developing-afghanistan-hydropower.

plants can bring about the support of entire communities against militants, while bringing water and electricity supply.[87]

Institutional Poverty

Decades of political instability in the country have led to the deterioration of the institutional structures across all sectors of the Afghan government and economy.[88] Due to increased political turmoil, an unstable economy, and increasing unemployment trends, the country continues to face an ever-growing brain-drain from its working-age population.[89] As a result, there is continued dependence on imported foreign expertise and inputs which is not sustainable in the long-run. Since 2014, international assistance has been greatly reduced, the only alternative to fill the consequent institutional voids with is by hiring local staff. However, there are ongoing issues with low staff capacity and shortage of critical expertise. Many staff lack awareness of developments in technologies which are being employed with success abroad. At the same time, complicated administrative procedures slow down the progress of the various institutions that are tasked with conducting research or exercising authority of water resources. The poor economic conditions of the country (i.e. GDP per capita, USD 585)[90] has hindered institutions such as government departments or university faculties from hiring, engaging, training and empowering technical staff from the international market in order to utilize their expertise for local skill-building. This is the key reason that public institutions in Afghanistan, dealing with water resource management or otherwise, perform relatively poorly as compared to those of the neighbouring countries.

From a legal point of view, the key positions of the water relevant ministries are not subject to hiring conditions based on technical know-how or

---

[87] Hessami, E. (2018, December 4). Power Play: Can Micro-Hydropower Electrify Remote Afghanistan and Promote Peace? *New Security Beat.* https://www.newsecuritybeat.org/2018/12/power-play-micro-hydropower-electrify-remote-afghanistan-promote-peace/.

[88] Ahlers, R., Brandimarte, L., Kleemans, I., & Sadat, S. H. (2014). Ambitious Development on Fragile Foundations: Criticalities of Current Large Dam Construction in Afghanistan. *Geoforum, 54,* 49–58.

[89] Hutchinson, B. (2016, May 12). Security and Brain Drain in Afghanistan. *Words in the Bucket.* https://www.wordsinthebucket.com/security-and-brain-drain-in-afghanistan.

[90] World Bank. https://data.worldbank.org/indicator/NY.GDP.PCAP.CD?locations=AF.

professional background. Instead, human resources decisions for key positions are made for political reasons, reflecting the relationships and power networks of those in power. As a result, such political machinations lead to the sacrifice of better understanding of the main priorities of these ministries. Therefore, due to politics there is limited scope for improving the chronic human resources shortcomings within the key ministries dealing with water governance.

The overlapping mandates of several parallel institutions working in the water resources sector i.e. Ministry of Agriculture, Irrigation and Livestock (MAIL), Ministry of Energy and Water (MEW) and Ministry of Rural Rehabilitation and Development (MRRD) has prevented the emergence of a clearer vision for an integrated water resources management strategy capable of meeting the demands of the Afghan population and economy from the resources available. The relevant institutional setup therefore requires to strategically invest in the development of irrigation infrastructure (e.g. canal expansions, water storage structures, etc.) by considering an integrated approach and demarking clearer working areas and achievable goals; it is also required to build the capacity of the technical manpower of the aforementioned institutions to allocate reasonable quota for the key crops during the peak demand period as well as fulfil the requirements for groundwater recharge, municipal, industrial and mining purposes which might be more demanding in the years ahead.[91]

Furthermore, the current higher education system has not been synchronized with the nation's water resource requirements for all of the major sectors i.e. agricultural, municipal and industrial. In recent years, there have been various hydrometeorological stations' installations made along the river networks of Afghanistan. However, due to a lack of technical capacity on the part of the relevant institutes operating these stations, these have been installed incorrectly or have failed altogether.[92]

Therefore, there is an ongoing issue with the lack of reliable data regarding water resources. What scant meteorological data is being collected is not sufficient for undertaking thorough hydrological analysis which are

---

[91] Akhtar, F., Awan, U. K., Tischbein, B., & Cheema, U. W. (2018). Assessment of Irrigation Performance in Large River Basins Under Data Scarce Environment—A Case of Kabul River Basin, Afghanistan. *Journal of Remote Sensing, 10*, 972.

[92] Akhtar, F. *Water Availability and Demand Analysis in the Kabul River Basin, Afghanistan* (PhD dissertation). http://hss.ulb.uni-bonn.de/2017/4824/4824.pdf.

required for producing responsive and adaptive water governance systems.[93] Given the growing population of the country, being able to meet the current and future demands of the agriculture, industry and municipal sectors is a clear priority. In order to achieve this, Afghanistan requires strategic planning and investment in the institutions concerned with water resources management sector.

Water Resources and Pollution
According to analysis in recent years, water pollution is one of the key compounding factors expected to cause degradation in regional food security, reduced resilience to floods and droughts, as well as increased regional tensions by 2040 in both the Amu Darya and Indus basins.[94] At present, both surface and groundwater pollution pose threats to environmental and human health in Afghanistan. Roughly 27% of Afghanistan's population has access to improved water resources, and this figure falls to 20% in rural areas which is the lowest percentage in the world.[95] Since 2010 waterborne diseases have caused more deaths of Afghans than conflict has; an estimated 17 million Afghans continue to drink water deemed unsafe for human consumption.[96] The country's capital, Kabul, hosts a population of 6 million, 80% of which has limited access to hygienically safe drinking water.[97] One of the many reasons for such appalling numbers is that water infrastructure has deteriorated in the past decades. Due to poor access to sanitation services, 20% of the population, mostly rural, practice open-air

[93] Ibid.

[94] US Department of Homeland Security. (2012). *Global Water Security: Intelligence Community Assessment*. Washington, DC: Intelligence Community. https://www.dni.gov/files/documents/Newsroom/Press%20Releases/ICA_Global%20Water%20Security.pdf. Accessed December 24, 2018.

[95] *Afghanistan's Water Crisis*. https://www.hydratelife.org/afghanistans-water-crisis/. Accessed December 12, 2018.

[96] Gupta, J. *Polluted Waters Bring Disease and Death to Afghans*. https://www.thethirdpole.net/en/2017/07/27/polluted-waters-bring-disease-and-death-to-afghans/. Accessed December 12, 2018.

[97] *Afghanistan's Water Crisis*. https://www.hydratelife.org/afghanistans-water-crisis/. Accessed December 12, 2018.

defecation,[98] discharge sewage to rivers which through downward percolation or seepage in lateral flows contaminate the groundwater, the dominant source of drinking water, across the country. An absence of wastewater treatment facilities in the country due to its financial limitations and a lack of technical institutional and human resources. Consequently, as alternative septic tanks are constructed, their leakage contaminates groundwater, and in turn, contaminates the groundwater wells which are being exploited for drinking purposes. Approximately 25% of deaths among children <5 are directly attributed to contaminated water and poor sanitation. A reported 54% of children aged 6 months to 5 years exhibit stunted growth caused by exposure to contaminated water, while 67% are underweight.[99] Another enduring health problem is the Helminth infection which is also caused by poor water quality, sanitation and hygiene.

In 2016 a survey of the water in Afghanistan's capital city, Kabul, revealed faecal coliform concentration was above 5 parts per 100 milliliters of water in 80% of the places measured. While according to the World Health Organization (WHO) guidelines, for safe drinking water, *E.Coli* should be so low as not to be detected in 100 milliliters of water.[100] Another study on groundwater quality in Kabul concluded that drinking water withdrawn from groundwater constitute a potential health risk from boron, magnesium, manganese and sodium may not be used in the long term.[101] If the Government of Afghanistan does not intervene in a timely manner, the health situation of the growing population will reach crisis proportions due to a lack of safe drinking water. To this end, there is a dire need for a sustainable, multisectoral strategy, safeguarded by effective and comprehensive laws and robust institutions.

---

[98] Ibid.

[99] *Afghanistan's Water Crisis.*

[100] Gupta, J. (2017, July 27). Polluted Waters Bring Disease and Death to Afghans. *Third Pole.* https://www.thethirdpole.net/en/2017/07/27/polluted-waters-bring-disease-and-death-to-afghans/.

[101] Sundem, L. Quality of Drinking Water in Afghanistan (Master thesis). https://brage.bibsys.no/xmlui/bitstream/handle/11250/278526/MSc%20Lise%20Sundem.pdf?sequence=1.

Reforming Water Resources Governance in an Insecure Domestic & Transboundary Context

Recent studies have concluded that, despite the demands on the resources of the Kabul River Basin for increasing for cultivation, pasture and urbanization, there is no effective integrated land or water use planning by the government which could allow for optimization and more sustainable outcomes.[102] Therefore, research and policy discourse have emphasized the need to enhance land and water productivity,[103] minimizing conveyance losses along irrigation canals, and increasing on-farm water application efficiency.[104] However, the ability of the Afghan government to do this has been hamstrung, not only by its 'weak state' context but by protracted conflict.

The 2009 *Water Law* enacted by the Afghan government, with international donor stewardship, was based on the principles of Integrated Water Resources Management (IWRM). The formulation of this law came about as a means by which to regulate water use through a formal system of permits, to replace the water management regime of adjudicating common contestations over water through formally, informally or tribally. While this approach has been seen as beneficial from the standpoint of sustainability and climate change resilience, it is also argued that future success in water resource governance in Afghanistan will be predicated on improving the policy and analytical capacities of stakeholders.[105]

The *Water Law* regulates ownership, fees, rights, permits and usage of this precious resource.[106] The constitutional articles (i.e. Articles 10–40)

---

[102] Najmuddin, O., Deng, X., & Siqi, J. (2017). Scenario Analysis of Land Use Change in Kabul River Basin—A River Basin with Rapid Socio-Economic Changes in Afghanistan. *Physics and Chemistry of the Earth, Parts A/B/C, 101*, 121–136.

[103] Rout, B. (2008). *The Way the Water Flows*. Kabul: Afghanistan Research and Evaluation Unit. https://areu.org.af/wp-content/uploads/2016/01/811E-Typology-of-Irrigation-Systems.pdf.

[104] For example, The World Bank. (2018). *AF On-Farm Water Management (OFWM)*. http://projects.worldbank.org/P120398/on-farm-water-management-ofwm?lang=en.

[105] Habib, H., Anceno, A. J., Fiddes, J., Beekma, J., Ilyuschenko, J. M., Nitivattananon, V., & Shipin, O. V. (2013). Jumpstarting Post-Conflict Strategic Water Resources Protection from a Changing Global Perspective: Gaps and Prospects in Afghanistan. *Journal of Environmental Management, 129*, 244–259.

[106] Wegerich, K. (2010). The Afghan Water Law: "A Legal Solution Foreign to Reality"? *Water International, 35*(3), 298–312.

related to water management emphasize the establishment of river basin agencies, water user groups and securing the rights of the water users in different reaches of the canals.[107] In addition to these river basin agencies, the Ministry of Agriculture, Irrigation and Livestock (MAIL), Ministry of Energy and Water (MEW), National Environmental Protection Agency (NEPA) and the Ministry of Rural Rehabilitation and Development (MRRD), etc. are empowered to regulate water resources at different administrative levels.[108] These management bodies are responsible for efficient and sustainable water resources management, equitable distribution, strengthening the national economy in accordance with the principles of Islamic jurisprudence and the local customs and traditions.[109]

However, there have been some concerns that the provisions of the *Water Law* are difficult to implement in traditional irrigation systems and serve international donors and their agenda to create a national hydrocracy more than local needs and conditions.[110] Since the aforementioned law was passed by the parliament, scant attention has been given to strengthening water resources relevant institutions or to addressing the lack of technical skills within these institutions. This has weakened the effectiveness of the implementation of the *Water Law* and has consequently limited the prospects of sustainable planning and efficient water resources management across the country. In turn, this has hampered the state's ability to respond to effectively to water resource crises in the form of frequent droughts and floods that have become a fixture across the country.

Furthermore, insecurity and 'warlordism' also prevent law enforcement agencies from governing the water resources effectively. It is also worth underscoring that one of the key reasons for mismanagement of water resources is the prevailing insecurity across the country.[111] Instability and insecurity have hampered accessibility on the part of state actors with jurisdiction over water resource management to different sites where water management stations and structures require construction. This situation

---

[107] *Water Law 2009* (Afg) a 2. http://www.cawater-info.net/bk/water_law/pdf/afghan_water_law_2009_e.pdf, November 30, 2018.

[108] Ibid.

[109] Ibid.

[110] Wegerich, K. (2010). The Afghan Water Law: "A Legal Solution Foreign to Reality"? *Water International, 35*(3), 298–312.

[111] Alim, A. K. (2006). Sustainability of Water Resources in Afghanistan. *Journal of Developments in Sustainable Agriculture, 1*(1), 53–66.

26   F. AKHTAR AND U. SHAH

has left many key projects and plans unfulfilled or delayed. Furthermore, insecurity allows for the perpetuation of an overall situation of lawlessness. This creates the conditions for water allocation and extraction rules and regulations to be violated,[112] especially by local power figures at the expense of water users from more marginalized sections of the population. In the desert areas of Helmand province that have seen increased levels of irrigation of poppy fields using groundwater, the government has effectively ceded control to the Taliban. However, it has been reported that the Taliban claim to have begun to monitor and regulate the over-extraction of groundwater.[113]

Aside from contestations over control of water resources by armed opposition groups such as the Taliban, a further problem that has plagued effective water resource governance in Afghanistan has been the constellation of parallel, and often competing, official administrative structures. These include the above-mentioned MAIL, MEW, NEPA and MRRD structures that each have their own competing agendas. This lack integrated governance for the water sector in Afghanistan has been held up as an impediment for sustainable development but has yet to see substantial improvement.[114]

Recent policy research has called for the establishment of an independent, technical and powerful central governing body for water resources.[115] This body would implement the central water management agenda and the various line ministries mentioned above would lose their authority over water governance. Such proposals are attractive as they could potentially see solve the issue of contested authority over resources that occurs between various institutions. The same research called for the water sector portfolios of line ministries to be restructured from the current administrative units to the natural boundaries of a given river.[116] In this spirit, President

---

[112]UNFCC. n.d. *Water Resources and Adaptation programs in Afghanistan (United Nations Framework Convention on Climate Change.* https://unfccc.int/sites/default/files/afghanistan.pdf.

[113]Mansfield, D. (2018). *High and Dry: Poppy Cultivation and the Future of Those Residing in the Former Desert Areas of South West Afghanistan.* Kabul: Afghanistan Research and Evaluation Unit, 17. See also, following section.

[114]King, M., & Sturtewagen, B. (2010). *Making the Most of Afghanistan's River Basins* (p. 11). New York: East-West Institute.

[115]Ahmadzai, A., Azizi, M. A., & Behzad, K. (2017). *The Impacts of Water Sector Reforms on Agricultural Productivity in Afghanistan.* Kabul: Afghanistan Research and Evaluation Unit.

[116]Ibid.

Ashraf Ghani has established a High Council on Land and Water. This body is at present formulating Afghanistan's first water sector policy, involving feasibility studies and surveys on the country's water resources.[117]

The transboundary nature of all but one of Afghanistan's river basins means that already challenging issues of internal water governance are ever more problematic as domestic demand continues to rise. The current river basin agencies are national in their present iteration and are limited in their capacity to contribute to transboundary developments without significant developments. Policymakers have evidently recognized the place of transboundary cooperation over shared water resources when the Ministry of Energy and Water (MEW) drafted the Water Sector Strategy.[118] But this has not translated into any substantive change and there is an ongoing lack of awareness among policymakers about the potential for internal contestations over authority over water resources to foment transboundary tensions as a result of such internal.[119] Some experts have called for the establishment of shared and transparent data systems for hydrological data on each of Afghanistan's transboundary rivers.[120] Such efforts could go a long way to bridging trust deficits that have plagued the country's attempts at regional economic development. That said, some research suggests that Afghanistan has been able to transcend its internal weaknesses, and take advantage of the international presence in the country and some of its neighbours' foreign policy constraints (Iran and Turkmenistan), in order to establish hegemony over the management of the Hari Rod basin.[121] Without transboundary trust-building efforts, however, such gains may simply lead to further contestation over water resources.

Therefore, establishing an effective system by which to govern water resources in Afghanistan, and lift the country towards self-reliance, is predicated on a transparent legal framework for water governance and a joint,

---

[117]Haidary, M. S. (2018, March 28). As Kabul Grows, Clean Water a Step Toward State Legitimacy in Afghanistan. *In Asia: Weekly Insights and Analysis.* https://asiafoundation.org/2018/03/28/kabul-grows-clean-water-step-toward-state-legitimacy-afghanistan/.

[118]King and Sturtewagen, 11.

[119]Ahmadzai, S., & McKinna, A. (2018). Afghanistan Electrical Energy and Trans-Boundary Water Systems Analyses: Challenges and Opportunities. *Energy Reports, 4,* 435–469.

[120]Ibid., 12–13.

[121]Thomas, V., & Warner, J. (2015). Hydropolitics in the Harirud/Tejen River Basin: Afghanistan as hydro-hegemon? *Water International, 40*(4), 593–613.

good faith effort by the country and its neighbours, with financial and scientific support from the international community. In conjunction with the Supreme Council governing water issues and a robust and effective water law, domestic water issues would for once be discussed in an institutional context which could allow for sustainable, efficient and equitable outcomes. As for the transboundary issues, some experts have called for the drafting of clear treaties with neighbouring countries, and updating the existing treaty with Iran, could advance better hydro-diplomacy in the region and mitigate potential conflicts as water resources grow increasingly scarce.[122]

[122]Hessami, E. (2018, July 11). As Afghanistan's Water Crisis Escalates, More Effective Water Governance Could Bolster Regional Stability. *New Security Beat*. https://www.newsecuritybeat.org/2018/07/afghanistans-water-crisis-escalates-effective-water-governance-bolster-regional-stability/; Akhtar, F. (2017). *Water Availability and Demand Analysis in the Kabul River Basin, Afghanistan* (PhD dissertation), University of Bonn.

Water Management in Bangladesh: Policy Interventions

Punam Pandey

Introduction

Although Bangladesh is not originally considered as part of Himalayan South Asia, it has been included in this study because of the reasons mentioned by the editor in the preface of this book. Theoretically, Bangladesh could have been one of the wettest places in the world considering the range of water bodies present in the country; in reality the availability of water differs temporally and spatially. Bangladesh suffers from the extremes. In the monsoon the surface water inflows go from maximum of about 14,000 cubic meters in August to a minimum of about 7000 cubic meters in the dry season of February. Spatially the amount of annual rainfall ranges from about 1600 mm in the southeast to about 3200 mm in the northeast region of the country. Managing such contrasts can be an uphill task for any country, it is definitely a challenge for a country that has to fight multiple challenges on many counts—burgeoning population to a greater need for economic prosperity—also have to tackle the vulnerability of the climate change phenomena.

P. Pandey (✉)
Research Associate, Institute for Reconciliation and Social Justice, University of the Free State, Bloenfontein, South Africa

© The Author(s) 2020
A. Ranjan (ed.), *Water Issues in Himalayan South Asia*,
https://doi.org/10.1007/978-981-32-9614-5_2

Bangladesh is a riverine country. Out of about 230 rivers, it shares 54 with India and three with Myanmar. Almost every year some part of Bangladesh suffers a flood and some other parts deficit. Though floods are beneficial for groundwater recharge, soil fertility, moisture and renewal of fish stock, it brings massive destruction as well time to time. This demands excessive mobilization of precious resources which the young nation needs to deliver public goods to people.

Interestingly, the centre of discussion on water issues in Bangladesh has mainly been interstate because of nature and extent of the country's dependence on neighbours. Though intertwined, closer examination brings out many facets of water management that need to be investigated thoroughly; they do pose difficult challenges to the well-being of people. The purpose of the chapter is to examine these aspects and analyse policy responses.

The first section presents the physical background of Bangladesh. It covers the aspects that include rainfall, hydro-morphology, land types and, also discuss water as an input in socioeconomic activities. The second section analyses the interconnection between internal and external dimensions of water management in the country. Next section highlights the overall importance of water in the ecosystem of Bangladesh. The last section deals with sustainability of these functions of water in the emerging climate change scenario and ends with exploration of possibilities for future challenges and other aspects of climate change. Thus the landscape of Bangladesh water management must be understood in the context of its geography, socio-economic aspirations and environmental factors.

Physical Context

Bangladesh's topography is formed by three of the largest river systems of the world. They are the Ganga, Brahmaputra and Meghna. Bangladesh is an important constituent of the Bengal basin which have been built over the period by the alluvial deposits, brought by these rivers from the Himalayas. The combined catchment of these rivers is about 1.7 million kilometer which extends over Bhutan, China, India and Nepal but Bangladesh makes up only 7–8% of the watershed.[1] Floodplains constitute about 80% of the

---

[1] Ahmad, Q. K., Biswas, A. K., Rangachari, R., & Sainju, M. M. (Eds.). (2001). *Ganges-Brahmaputra-Meghna Region: A Framework for Sustainable Development*. Dhaka: University Press Limited.

total land area while higher lands account for about 8 and 12%, respectively.[2] The altitude normally does not exceed 11 meter above the sea level except in the hilly areas of Chittagong and Sylhet.

Bangladesh is one of the most densely populated countries in the world. Being a deltaic, it is a plain area; low-lying and mostly flat. The alluvial aquifer system of Bangladesh is one of the most productive groundwater resources in the region. A network of about 230 rivers forms a web of interconnecting channels throughout the country; apart from the surface water resources, the country has some other varieties of static water bodies such as ponds, hoars and others. All these sources of water keep the country's crop fields and other vegetations lush green throughout the year.

Bangladesh's climate is subtropical, as true to its characteristics, this brings intense rainfall in a limited time. More than three-fourth of the rain takes place in three months of July, August and September. Apart from flood, water logging is a regular feature of the country in these months of the year. Rest of the year witnesses scant to no rain in some places; thus annual rainfall varies from 1200 mm in the north-west to 400 mm in the north-east.

The surface water inflows vary from maximum of about 14,000 cubic meters in the months of August to a minimum of about 7000 cubic meters in the dry season of February. This vast gap poses a challenge to the governance because the nature of topography is not allowing the country to harness excess water to its advantage. The two significant upstream rivers like the Brahmaputra and Ganga originate outside Bangladesh, which lead to many other complications.

Nature and quantity of availability of water determines to a large extent the kind of economy the country pursues. The next section deals with economic requirements of water and realities of availability.

Nature of Economy: Being an agrarian economy, farming uses most of the water. Agriculture contributes handsomely to overall economy of the country and more than half of the population is employed in this directly or indirectly. Obviously large tracts of land are used for cultivation. At present, more than 50% of land is supported by some sort of irrigation facilities. On the one hand, demand for irrigation is growing, on the other hand, as discussed previously Bangladesh's water resources—surface as well

---

[2] Hasan, S., & Mulamoottil, G. (1994). Natural Resource Management in Bangladesh. *Ambio, 23*(2), 141–145.

as groundwater—have been signalling signs of stress; alternative model of farming with less water has not been developed or encouraged yet.

Though majority of population live in rural areas, Bangladesh is transitioning even somewhat slowly to urbanization. It is expected that by 2025, almost half of Bangladeshis will be living in urban areas. This upward mobility presents its own challenges, changes in life style demand household with comfortable appliances which suck more water. This highlights the growing appetite for more assured water supply in Bangladesh. This is about emerging and growing middle class. The poor have different sets of needs and demands.

Bangladesh has always had various schemes to provide safe drinking water to the people due to active civil societies presence in governance.[3] But for many people getting water delivered at home is still a dream.

Twining of Internal and External River Systems

As discussed above, Bangladesh lies at the downstream end of the river basins of three large river systems-the Ganga, the Brahmaputra and the Meghna. They form the largest riverine delta in the world. Bangladesh shares the largest number of rivers with India and a few with Myanmar.

Most important river for Bangladesh in terms of volume is the Brahmaputra. It rises in on the northern part of the Himalayas. Out of the total catchment area, less than ten per cent lies in Bangladesh. As with other Himalayan rivers, snowmelt contributes largely to replenish the flow before the monsoon. But the river picks in July and mid-September due to the Monsoon rain. The dry period is supported by melting of the snow. The river's salient feature is movement of high volume of sediments consisting of sand, silt and clay.

These deposits are laid on water bodies and raise their bed which lead to flood even in the case of modest rain. Rising of bed leads to reduced capacity of the rivers to absorb excess water. At the same time, these rivers are more susceptible to river bank erosion. Environment and GIS (EGIS) Project studied Brahmaputra-Jamuna in 1997 which made an observation that the average width of the river has increased by about 130 meter per year since 1973 and appears to have been slowly widening since that time. Riverbanks are both retreating and eroding the floodplain.

---

[3] Chowdhury, N. T. (2010). Water Management in Bangladesh: An Analytical Review. *Water Policy, 12*, 32–51.

Bangladesh faces serious water security concerns as it is always at the receiving end of the upstream river practices of India, Nepal and China. Erosion's impact on the country can be underlined in terms of flood protective measures and the potential ecological damage to the downstream regions.

Another important river for Bangladesh is the Ganga. The river originates in the southern slopes of the Himalayas in India and moves in a south-east direction towards Bangladesh. The largest portion of the Ganga lies in India. The Ganga is known as Padma in Bangladesh. The river joins Brahmaputra in the heart of Bangladesh and their joint flow then runs south to merge into the Bay of Bengal. Out of the total drainage area of the Ganga, 79% belongs to India, 4.2% to Bangladesh and almost 14% to Nepal.[4] India constructed the Farakka barrage on the Ganga near Bangladesh border in 1976. Of late, many structures have been constructed upstream of the Ganga which impede flow of water reaching to farakka. These structures are mainly for irrigation purposes. Water availability below the barrage somewhat declines dramatically during the dry season.

Among the trio, Meghna originates in hills of Manipur in India. The river flows between India and Bangladesh. As many other rivers in Bangladesh, Meghna is also rain-fed.

It is important to understand here that, of the total annual flows of Bangladesh, about 67% is contributed by the Brahmaputra, 18% by the Ganga, and about 15% by the Meghna and other minor rivers.[5] The contrast is more in the case of the Ganga; the ration between the discharges of the dry and monsoon seasons for the Ganga is 1:6.[6]

Issues concerning these rivers: the above discussion underlines the interconnection between river sharing countries of India, Bangladesh and China. But negative impact is felt harshly by Bangladesh because of its topography. The following section highlights the multifarious impacts on Bangladesh.

---

[4]Chellaney, B. (2011). Exploring the Riparian Advantage: A Key Test Case. In *Water: Asia's New Battleground* (p. 172). New Delhi: HarperCollins Publishers India with India Today Group.

[5]Ahmad et al. (2001: 37).

[6]World Bank. (2000). *Bangladesh: Climate Change and Sustainable Development* (Bangladesh Report No. 21104-BD), Rural Development Unit, South Asia Region, Document of the World Bank.

Recently while the Ganga and Brahmaputra have seen high level of sediments, the presence of organic matter has gone down. The presence of organic matters is considered good for the fertility of a land. The result is gully erosion, further intensified by the removal of the vegetative cover in the upper catchments in India, China and Nepal.

Sedimentation: As Himalayan rivers make a long journey from Himalayas from Tibet in China through India and Nepal to Bangladesh, they pick up loads of sand, clay and silt, they unload them in the lowest stretch of the delta. The volume of these sediments has recently gone up high because of deforestation in higher lands due to road construction and other infrastructure projects. These heavy loads are gradually changing topography which in turn reduce the carrying capacity and navigability of various routes of rivers. The situation becomes challenging in the post-monsoon period when river flow goes down. This threatens to hitherto management practices and ecological health of waterbodies.

Watershed Management

As discussed earlier, Bangladesh receives almost 93% of river water from outside its territory. It is corollary that health of watershed in India, Nepal, China and Bhutan determine the recharge of surface and sub-surface water of Bangladesh. Due to it, development especially in agriculture field is depended on water supply from these countries. Since infrastructure development in these countries have speeded up, it is causing adverse impact on environment. As for an illustration, India is actively pursuing construction of roads and infrastructure projects along border areas with China. Though this might be considered good for strategic and geopolitics, it is shaking the ecosystem of the region. At the same time, encouragement of tourist flow in Nepal has driven development in the upper sections. Forests are being cut to pave way for these projects. Deforestation is contributing to loosening of soil in upland areas of India and Nepal. Thus the result is massive soil erosion. Even in Bangladesh to promote tourism and develop infrastructure deforestation is being carried out. These actions do have impact on the nature.

Ganga serves more than 500 million population in South Asia. Before entering Bangladesh, it traverses heavily populated and agriculture dependent provinces in India. Heavy Population as well as agricultural requirements have compelled the government to construct structures for diversion of water. Excessive withdrawal of water from upstream obviously leads to adverse impact on downstream neighbour.[7] As agriculture is the means of livelihood for millions of people, deficit of water impact two ways—salinization of agricultural land as well as lack of assured water for farming. There are interconnections between aquifers across the region, whatever action is taking place in one section of the river and aquifers, the impact is also going to be felt in the other end as well.

Spatial and Temporal Dimensions of Water

Spatially and temporally, availability of water varies a lot. During the dry season (November to May), there is a serious shortage of water amidst high demands, it requires deft management of water amidst many competing needs across many sectors and levels. During the monsoon, surface water is available in excess of demands. This gives little advantage because of flat topography. Circumventing this physical barrier remains a big challenge.

Apart from rivers, a network of channels and other water bodies covers 8.23% of the surface area of Bangladesh. As in other parts of South Asia, the country is also suffering from depletion of groundwater. Rainfall and annual inundation—a normal feature for Bangladesh—are not able to cover the deficit and replenish the groundwater.

Water Usage in Socio-Economic Activities

Assured supply of water plays a critical role in scaling economy as well as overall development of the country. Apart from being an important input for agriculture, water plays a critical role in many other areas like fisheries, navigation. In Bangladesh, agriculture makes 22% of GDP and employs about more than half of the country's labour force.[8] Agricultural sector is the major consumer of water. For Bangladesh, the world's fourth biggest producer of rice is a very important crop for dietary habits. Rice production is one of the main sources of revenue for the country and one of the important export items. Crop cultivation has been scheduled according to rain and temperature phenomena—pre-monsoon, monsoon

---

[7] Khuda, Z. R. M. M. (2001). *Environmental Degradation Challenges of the 21st Century.* Dhaka, Bangladesh: Environmental Survey and Research Unit.

[8] Government of Bangladesh. (2005). *Unlocking the Potential National Strategy for Accelerated Poverty Reduction.* General Economics Division, Planning Commission and Government of the People's Republic of Bangladesh, Dhaka, Bangladesh.

36 P. PANDEY

and winter season. Aman (rain fed rice) is the main rice crop that covers more than 50% of the total rice grown area. But of late, boro (dry season rice) is becoming preferred choice among farmers; this variety is heavily supported by irrigation.[9] Another related economic activity is fishery.

Fisheries: Bangladesh is blessed with one of the largest inland fisheries in South Asia. Hilsa is the most prized possession of the country. Fish constitutes one of the important component of the protein intake. It is also an important source of employment and contributes about 6% of the total GDP.[10]

Navigation: Water is an important mode of transport especially for weaker sections of the society. Overall transport accounts for 8% of GDP and water transport accounts for about 15% of total transport GDP.[11] About 30% of all national freight and 14% of the passengers use inland waterways.[12] Boats are now mainly prevalent in rural areas; but they can still be sighted carrying people in certain areas even in the capital city, Dhaka. It is estimated that there are more than 8 million in the country boats that can carry traffic 20 times the capacity of trucks in Bangladesh. This mode of transportation becomes important for bulk goods; more significant for being environment friendly.[13]

Industry: Though Bangladesh is a predominantly agricultural economy, industrial activity especially related to textile is picking up. As with other parts of South Asia, industrial and domestic demands of water is miniscule in comparison to agricultural requirements; but they have disproportionate contribution towards pollution.

In industrial sector, water is an important input in the production and also used for cooling purposes. There are some industries which use water directly as an input, as for an example, paper and pulp industries, newspaper mills. Some industries get water through municipal connections; others use surface as well as groundwater directly. They use water at their will.

---

[9] Water Resources Planning Organisation (WARPO). (1999). *National Water Management Plan.* WARPO, Dhaka, Bangladesh; Ahmad et al. (2001).

[10] World Bank. (2006). *Bangladesh Country Water Resources Assistance Strategy* (Report No. 32312-BD). Environment and Social Development Unit, Agriculture and Rural Development Unit, Energy and Infrastructure Unit, South Asia Region, Document of the World Bank.

[11] Ibid.

[12] Ibid.

[13] World Bank (2000).

Importance of Water in Biodiversity and Environmental Sustainability

Bangladesh hosts a great variety of flora and fauna that change with diverse climatic regimes, water availability, land types and so on. Seasonally over 60% of Bangladesh can be classified as wetlands.[14] Wetlands serve the hydrological functions of maintaining the subsurface water table through recharge of aquifers and act as a buffer against flood. Together with neighbouring India, Bangladesh supports one of the largest mangrove forests in the world-the Sundarbans. The Sundarban protects the region from storm and cyclones. This forest is home to Bengal Tigers and other animal species.

Externality of Upswing of Industrial Sector: In order to achieve the middle income group, Bangladesh has been trying to diversify the economy. So industrial activity has been incentivized. Behind the phenomenal economic growth of Bangladesh in recent times, the textile sector has played a tremendous role; this has contributed more than three-fourth of the country' total export earnings. But this success has come with a price. This is proving to be adversarial for the environment especially in the water sector. Many textile factories withdraw water from their own property and treat them as free commodity. This has led to an unwise use of water resources. Since some of the garment factories are water intensive, they do not dispose of polluted water in a safe way or recycle them. Thermal power plants and other industries also use water for cooling purposes. They are adversely impacting the environment.

Though Bangladesh gets more water per capita, its hydropower generation and storage of surface water is constrained because of Bangladesh's flat surface and high population density. Kaptai dam is the only notable hydropower facility in the country. The upstream neighbours of shared rivers have high potential for hydropower generation but these measures have been caught in geopolitics of the regions. So, each country is battling with diminishing water resources in its own way.

WATER MANAGEMENT CHALLENGES

The above discussion suggests the abundance of water in monsoon and dire stress in summer. Thus the optimal utilization becomes necessary as

---

[14]World Bank (2006).

38 P. PANDEY

well as urgent. Department of Environment (DoE) has been monitoring surface water quality since it is established in 1973. Some of the important challenges faced by Bangladesh are analysed below.

Flood is an annual feature in Bangladesh but some of them have been really devastating which has caused a great loss of life and livelihood as for an example in 1988, 1998 and 2004. In this kind of scenario, it takes days to drain water from flooded areas. There has also been a lack of proper network of infrastructure in the sewage management.

Bangladesh Water Development Board (BWDB) has always prioritized Development of flood control drainage irrigation projects (FCDI) as part of water management in the last five decades. But intensive irrigation projects have not been able to accommodate drainage aspect in its planning which lead to water logging. FCDI projects can be structured as integrated water resource management that should contain aspects like navigability, fisheries and environmental issues. Embankments can be designed to have structures for controlling flood of major rivers such as the Ganga, Meghna and other smaller rivers. Non-structural measures include upgrading methods of flood forecasting and warning, flood proofing, disaster management and floodplain zoning.

Since Bangladesh suffers from extreme water shortage in dry season, the country has been looking for augmenting water supply. The country has approached India several times (1972, 1974, 1985) but both neighbours have not been able to agree about the ways to go about augmentation. In 1996, finally when the Ganga treaty was signed between the countries, the augmentation issues were not included explicitly in the text, both countries only expressed the intention to explore the options for augmentation. The BWDE and the Water Resources Planning Organisation (WARPO) have recommended a Ganga Barrage project.[15] This proposed project, if implemented, will address dry season's needs especially in South-Western region of the country. Environmentalists are not enthusiastic about the project.

In recent past, much of the way the Ganga traverses have been observed to be affected by the arsenic problem—colourless, tasteless and odourless substance found in millions of shallow wells across Bangladesh. Long-term exposure leads to serious health issues. About 20 million Bangladeshis, or one in eight, have been drinking water with arsenic levels higher than the government's limit of 50 microgrammes per litre. This is five times

---

[15] Government of Bangladesh (2005).

higher than the World Health Organisation (WHO) guideline of 10 microgrammes per litre. This is a severe problem in parts of Bihar, West Bengal in India and, this is affecting more than three-fourth districts of Bangladesh. Other rivers are also suffering from this.

The highest level of contamination is found in the large basins of the rivers Padma, Atrai, Brahmaputra and Meghna in the northern and the north-eastern part.[16] There would be high implications for food safety if irrigation continues with contaminated water. The solution seems to be in the treatment of tube wells or in collecting water from alternative sources such as surface or rain water.

Another important challenge related to water management is land erosion and accretion of rivers. Rivers are very dynamic because of young nature of Himalayas. Erosion of river banks is a normal feature in the monsoon and post-monsoon period. Erosion is also very common across all prominent rivers of the GBM. This leads to huge loss of agricultural land and human dwellings. In lower part of delta, tidal currents and storm surges from the sea lead to river bank erosion.

Moreover accretion of silted substances on rivers are not in the same proportion as erosion, it is way too high in some places. Controlling this problem in major rivers are expensive and risky as well. However, the government has initiated some projects for the purpose like a Jamuna Meghna Erosion Mitigation Project.

Apart from quantity, quality is a serious issue as well. The quality concern is severe in many places because of adoption of high yielding varieties, untreated domestic sewage and industrial residue. In fact, rivers have become dumping grounds for municipal solid wastes in some places. These polluting agents are causing serious harm to ground, surface and marine ecosystem. The surface water system is under heavy contamination due to indiscriminate discharge of industrial effluent and domestic waste water. The Pollution of closed water bodies caused by human waste is another major problem. Other concerns which need serious considerations are faecal contaminations in urban areas, village ponds and small streams.

Salinity intrusion occurs in both surface and groundwater aquifers. The effect of saline water intrusion is highly seasonal. Saline intrusion is at its minimum during the monsoon when the GBM rivers are at its peak. In winter months, the saline front begins to penetrate inland; this affects more

---

[16] Khuda (2001).

than one-fourth of areas in the dry period. Saline intrusion has emerged a big concern in the northeast region of Sundarbans. As a result, groundwater is suffering and becoming unsuitable for agriculture, industrial and human consumption. Salinity intrusion in surface and groundwater has also crippling impact in South West region of Bangladesh. The water supply of the area is mostly dependent on the Ganga.

Another consequence of pollution is that waterborne diseases are eroding the benefit gained by Bangladesh in the health sector. As discussed above, water quality is being compromised from various sources that includes arsenic, untreated sewage, seeping of chemical fertilizer and pesticide. Waterborne diseases are high in flood season because of inundation of hand pumps and latrines. As for an example, waterborne diseases caused by the floods of 1988 and 1998 resulted in a large number of causalities.

As discussed above fish is an important component of Bangladeshis' food. But fisheries depend on sufficient presence of water bodies and connections with migratory routes. But the decision to construct FCDI projects to prevent or regulate floods in paddy field has restricted the ability of migratory fishes to move across water bodies. This has led to tension between farmers and fishermen. These factors are contributing to reduced fish productivity and species diversity in many areas.

Another noticeable impact of water pollution has been in the biodiversity. Wetlands and different ecosystem have lost connections with other water bodies like rivers and canals due to siltation and filling of water bodies for agriculture and dwelling purposes. Wetlands have been particularly subject to particular pressure due to increasing dependence on irrigation. The degradation has brought about a loss of diversity, reduction in fish varieties and habitat; large swathe of land becoming infertile. Denudation of forests have affected availability of food, fodder, medicine and shelter for the poor people; access to basic amenities are very restricted to forest dwellers. Though pollution affects everyone but the hardest hit are the poor people.

Like many other countries of South Asia, agriculture and fishery are interconnected especially in the rice plantation areas. Now both seem to be in conflict because of focus on intensive agriculture and population pressure. River embankments have been built to protect dry and wet-season crops from flooding. This intervention has damaged natural fish habitats and migration routes.

Like India, Bangladesh is also suffering from extreme groundwater deficit. Exploitation has led to dryness of aquifers, no effective measures have been injected to replenish these aquifers. Bangladesh's capital city, Dhaka, is also reeling from extreme dryness.

The portentous signs and emerging realities of Climate Change are going to pressurize the already compromised water resources. It is expected that temperature of Ganga–Brahmaputra-Meghna river basins would increase up to 5.5% by 2020. This becomes particularly alarming for Bangladesh because the country is vulnerable to environmental hazard, in case of any extreme environmental incident, the country would experience disproportionate damage because of high density of population.

The glaciers of the Himalayan mountain ranges are the life-line of Asia's water supply. As discussed above, millions of people depend on these rivers for life-support system. Unlike monsoonal rain, these glaciers release water slowly into the rivers of the Indus, Ganga and Brahmaputra basins. The major contribution to the annual flow of the Ganga comes from the Nepal-Himalaya rivers, principally, the Karnali (Ghaghara), Gandak and Kosi. They are snow-fed and provide almost half of its lean season flow.[17] The climatic research forecasts that Himalayan glaciers are receding faster than expected previously. The emerging future appears alarming to people of not only Bangladesh but most of South Asians because the Ganga is heavily dependent on these glaciers for as much as three-fourth of its summer flow. The recent data shows that there is a regression of approximately 30 days in the maximum spring flow period and an increase of 30–38% in the glacial runoff. When the shortage arrives in a decade or two, it will be quite abrupt, as the flow will be dramatically reduced in the dry season.[18]

Another important concern related to climate change phenomena is about irrelevance of past templates for forecasting weather patterns. Past hydrological experiences are generally studied to provide a model to understand future possibilities and recommend the future preparedness accordingly. Till now, the dry season water availability in the rivers depend on the intensity of the previous monsoon rain. In case of any substantial change in the hydrological cycle of mountains, this would definitely alter the pattern of rainfall and consequently all countries including India and Bangladesh be immensely damaged. In fact, there are records which suggest rainfall

---

[17] Verghese, B. G. (1999). Water of Hope. *Waters of Hope from Vision to Reality in Himalaya-Ganges Development Cooperation* (p. 336). New Delhi: Oxford University Press.

[18] Pachauri, R. K. (2009, April). Climate Change and Water. *Asia's Next Challenge: Securing the Region's Water Future* (A Report by the Leadership Group on Water Security in Asia), Asia Society, p. 33.

pattern has already shifted substantially in the last one decade. There have been wild fluctuations in the seasonal rainfall, the timing, setting up, its peak and withdrawal of rainfall. Thus, many water resources, especially drinking water is increasingly threatened in both urban and rural areas.[19]

Climate Change is going to have debilitating impacts on Bangladesh in many other ways. The country is already prone to climate-related incidents such as floods, cyclones, storm surges, flash floods, droughts, river bank erosion and rain induced landslides. Bigger destabilization would be experienced in salinization of surface and groundwater. In fact, the Bangladesh government has admitted that out of close to hundreds of rivers, one has already died; another 100 are drying up. Drying up of rivers are affecting navigation too.[20] Some other parts of the country suffer from water logging. Every year almost one-fourth of the country suffer from inundation during the monsoon. In fact the frequency of terrible floods is going up which cover three fourth of the country. From 1970 to 2008, 12 major cyclones have killed more than 6 million people and affected other 45 million.

River erosion is a serious issue affecting the Ganga, Jamuna and other important rivers of Bangladesh. It affects not only people but also devour property and infrastructure to a large extent. It has been observed that large areas of land have been submerged, and rendered millions of people homeless. Erosion is not a new phenomenon but recently its frequency has gone up.

Some of the rivers are on the verge of extinction. The Buriganga, once considered a vital part of Dhaka, was a main source of drinking water for almost more than 20 million inhabitants of city of Dhaka. This also used to perform many important functions like transportation, cleaning, washing, groundwater recharge, flood control apart from having ecological significance. But the river has succumbed to chemical wastes generated by mills, factories, households, medical establishments, sewage, dead animals and plastics. According to the DoE, 22,000 cubic liters of toxic waste are released into the river by the tanneries every day. Textile industries annually discharge as much as 56 million tons of water and 0.5 million tons of

---

[19] Chowdhury, M. R. *Water: The New Oil.* http://www.thedailystar.net/newDesign/print_news.php?nid=213847. Accessed December 13, 2011.

[20] http://www.thedailystar.net/newDesign/news-details.php?nid=222372. Accessed February 14, 2012.

sludge: most of these are released in the Buriganga.[21] Most of the river bed is also lost to encroachment; municipal and factory waste dumped through storm drains.

Industrial pollution accounts for 60% of pollution in the Dhaka watershed area, and the textile industry is the second largest contributor after tanneries. Almost 3000 garment factories are operating in Dhaka. Industrial water and effluents containing heavy metals are being released in the vicinity of industrial areas and this polluted river water is used for irrigation purpose of paddy and vegetable near industrial areas in Gazipur. Untreated effluent is discharged into rivers from nearby textile factories like Gazipur, Tongi, Savar and Ashulia.

Lack of assured water Supply and diminishing surface water resources prompted the government authorities to install a groundwater supply system in the 1970s. Under this provision, a number of wells were dug up for providing safe drinking water. Presently, Bangladesh has more than 9 million wells. Though wells made it possible for about 97% of the rural population to have access to bacteriologically safe water by 2000 and helped lower the infant mortality rate from 156 per thousand in 1990 to 69 per thousand in 2006.[22] Unfortunately, particularly in shallow aquifers, the groundwater often contains arsenic at levels that can cause poisoning (arsenicosis). This naturally occurring arsenic is a major concern for drinking water supply, animal husbandry and irrigation. It is also a major development constraint in coastal aquifers. In 61 of the country's 64 districts, groundwater arsenic levels are above the permissible limit. It is estimated that almost 25–35 million people depend on wells that expose them to the risk of arsenicosis.

Bangladesh faces the maximum threat to the Brahmaputra from the actions of its two upper riparian neighbours as its population is dependent on this river water largely. China has been constructing and planning massive water infrastructure on the river which would reduce the water flow to Bangladesh. Though both India and China share seasonal water flow and rainfall data with Bangladesh to help with flood forecasting; Bangladesh

---

[21] Kibria, M. G., Kadir, M. N., & Alam, S. (2015, December). *Buriganga River Pollution: Its Causes and Impacts*. Conference Paper. https://www.researchgate.net/publication/287759957_Buriganga_River_Pollution_Its_Causes_and_Impacts. Accessed Research Gate, June 6, 2019.

[22] Bangladesh: The Confluence of the Ganges, Brahmaputra and Meghna Rivers. *Facing the Challenges*, The United Nations World Water Development Report 3 Case Studies Volume, London: UNESCO, 2009.

needs to make more emphatic presence felt in the development of the river.

Most of Bangladesh is on low level from sea, some places only inches or centimeter above sea. Massive Destruction of mangrove forests is being documented in both sides of Sundarbans in India and Bangladesh. The northern part of the Sundarbans is affected by salinization. Another climatic impact is the submergence of large tracts of land. This problem becomes more pronounced in Bangladesh because of higher possibility of flood as a result of drainage congestion and saturation. Immediate consequence is in increased number of days of water inundation and slow drainage.

One of the effective means suggested for removing mud and soil from large rivers is the use of dredgers, mainly power-operated heavy-duty dredges mounted on ships or boats; the running cost, however, is prohibitive. For comparatively smaller rivers, such as Ichhamati, Bangladesh could rely on the labour force of the villagers to undertake manual dredging during the dry season. The country could employ thousands of villagers for this purpose, based on the principle of food-for-work (FFW). Thus Bangladesh has to take all these challenges seriously, policy interventions have been made from time to time.

Policy Interventions

As Bangladesh's water woes are growing, the government has initiated many policies to address the situation. Bangladesh has tried to evolve a participatory approach to deal with water management issues. State agencies involve multiple stakeholders like NGOs, farmers, public, industry, sanitation, water and sewage, public health and municipalities, inland water transport, fisheries, forestry and environment. This participatory approach has also certain downside.

More than three dozen central government institutions affiliated with 13 different ministries, have responsibilities and activities relevant to the water sector. At top of the hierarchy, the Ministry of Water Resources is responsible for a number of activities—flood management, irrigation, drainage control, erosion protection, land reclamation, integrated management of coastal polders, river flow augmentation, water sharing on transboundary rivers and wetland conservation through participation of local people, and also coordinate programmes with all the other ministries dependent on water resources.

The National Water Resources Council (NWRC) is the highest national body for the formulation of water policy. It coordinates different water agencies and makes recommendations on all water policy issues to the cabinet. The National Water Policy formulated in 1999, has guidelines for agriculture, fisheries, industry, navigation, environment, basin-wide planning. The policy presents the broad contours of water resource development; its particular emphasis is rational utilization. The mainstay of planning is the importance of conjunctive use of ground and surface water.

All water sector projects need to conform to DoE rules and guidelines. It endorses set of Environment Impact Assessment (EMI) guidelines; this is executed by the Ministry of Water Resources. Most of the water Ministry projects are subject to DoE scrutiny when they are submitted to the Planning Commission. Close investigations suggest that WARPO, NWRC, the Ministry of Environment and Forestry and DoE all have to some extent overlapping roles.

As for an illustration, it has been discussed in previous pages that structures built to control floods in one area have detrimental impact on other sectors. Given the differential needs of various water management issues, improved coordination among these institutes would lead to better designed projects and achieve large economies of scale.

The National Water Policy (NWPo) published in 1999 aims for a holistic, multi-sectoral approach to water resource management. In 2001, the government introduced a National Water Management Plan, prepared by Water Resource Planning Organisation (WARPO). The plan's aim is to implement NWPo directives and decentralize water sector management. It provides a framework within which line agencies and other organizations are expected to coordinate planning and implementation of their activities.

An increasing demand for both surface and groundwater comes from irrigation. It accounts for 58.6% of total demand of water. However, in setting priorities for allocating water during critical periods, the National Water Policy 1999 gives this sector a relatively low priority. The Water Policy sets the priorities in the following order: domestic and municipal uses (e.g. navigation, fisheries and wildlife), river regime sustenance and other consumptive and non-consumptive uses including irrigation, industry, environment, salinity management and recreation[23]: Fisheries; navigation and environment sectors demand 40.7% while demand for household and industrial use is 0.7%.

Parliament has also passed the Water Act in 2013. This is the latest and the updated version of previous water regulations. The Act specifically recommends the formation of the high-powered NWRC with the Prime Minister as its head. This implies the higher level of importance the government is bestowing on management of resources. An Executive Committee under the Ministry of Water Resources would implement the decisions taken by the Council. According to the Act, all forms of water present in Bangladesh—surface, ground, sea, rain, atmospheric—belong to the government on behalf of the people. Thus the government has been entrusted the responsibility to protect water. Another remarkable feature is that it suggests initiative for a basin level integrated water resources management of transboundary rivers and also recommend exchange of data on flooding, drought and pollution with co-riparian countries.

These policy changes have underlined various aspects of water requirement and sets operating principles to fulfil varied functions of water. Since Bangladesh is dependent on neighbours for quality as well as sufficient quantity of water; the 2013 Act recognizes the reality of transboundary nature of rivers, it is instructive to explore opportunities with them to devise environmentally sustainable principles to guide water. Most importantly, the Management of water resources like rivers, creeks, resources, flood flow zone and wetlands has been assigned to the Executive Committee under the Ministry of Water Resources. Drainage of wetlands that support migratory birds has been prohibited.

The Act has provisions for punishment and financial penalty for non-compliance with the Act. There are broad categories of penalty for non-compliance that include negligence to abide by government policy, ordinance, non-cooperation with government officials, refusal to present necessary documents, providing false information, affiliation with perpetrators. Some of the provisions sound draconian and ham-handed.

Activists, civil society groups and environmentalists have been knocking the door of the court for restoring the health of rivers and other water bodies. This year 2019 the court has ordered on those petitions and asked

---

[23] WARPO (1999).

the government to make some significant changes to implement the court orders. Some of the court judgements are landmark. The judgement has touched many dimensions of river protection and involved many institutions in the country ranging from election commission to banks.

The court declared rivers as legal entity and made the National River Protection Commission (NRPC) as the legal guardian to act as their parents in protecting the rights of waterbodies, canals, beels, shorelines, hills and forests. Interestingly, the court made an observation that the NRPC should not just report to government but be empowered to take action suo moto. The government was also directed to amend the NRPC to make it more effective and independent.[24]

In fact the court has directed the Ministry of education that students should be given an hour-long class once in two months in all government and private academic institutions including schools, colleges, universities, madrasahs about the rivers. The Ministry of Industries is also requested to hold one-hour meetings in every two months among the labourers of mills and factories across the country for raising awareness regarding the protection of rivers.

Grabbers have taken advantage of dying of rivers due to low flow during lean seasons, and diminishing upstream flow in transboundary rivers to encroach on them. Many of the land grabbers are often sheltered by influential political and social groups in the country. The court observed that land-grabbers should be debarred from contesting election or seeking bank loans. So the encroachers would be disqualified from contesting elections for Union Parishad, upzila, Pourasava, city corporation and Jatiya Sansad election. The Bangladesh Bank has been advised not to grant loans to individuals or corporations found guilty of these acts. The court has also instructed the government to compile and publish a list of land grabbers. These are phenomenal features of the judgement which are path breaking in nature and set a precedent for environmental awareness.

[24] Rivers Are Now 'Legal Persons', *The Daily Star*, February 5, 2019. https://www.thedailystar.net/environment/news/can-the-historic-high-court-judgement-save-our-rivers-1697419. Accessed June 6, 2019.

Scope for Partnership to Fight Emerging Challenges

Climate change is driving every country to reimagine and redesign the structure and policies to deal with uncertain future scenario of water availability. Bangladesh has become conscious of climate change phenomenon and trying to address the issue in its own way by bringing in changes in its various policies. To address the fluctuation in the areas served largely by the Ganga water Bangladesh has decided to construct the Ganga barrage. The government was supposed to begin construction of the Ganga Barrage in December 2012 to preserve the river water during monsoon and release it during the lean period. But this construction work has been halted because of technical fault with the design of the barrage.

As discussed above, supply and health of Bangladesh's water is interlinked with its neighbours, any preservation plan initiated by India for Ganga would have positive externality on its eastern neighbour. In an attempt to address the new climatic challenges, the government of India has designed new water policies. This is important as the focus is to document the availability of ground and surface water resources in each river basin. This is also designed to study the health of aquifers as at basic level at watershed all over the country. Unlike present system of farming with uniform patterns, adaptation of these new processes would encourage farming according to agro-climatic conditions. At the moment flood irrigation is adopted and preferred by farmers, this leads to wastage of water and problem of salinization. Farmers, especially the big ones opt for crops according to commercial considerations. The government is incentivizing farmers to learn new techniques which respect the nature and availability of local water and soil conditions. This would definitely have a positive extrapolation to a riverine neighbour. Naturally, this would leave more water to flow to the farakka barrage and as an end destination to Bangladesh.

But recently, the World Bank assessment has reported that the Ganga river basin could see crop failures rise three-fold and drinking water shortage go up by as much as 39% in some places between now and 2040.[25]

---

[25] Koshy, J. (2019, February 24). Ganga Basin States Stare at Three-Fold Rise in Crop Failures by 2040. *The Hindu*.

The report further warned that 'the volume of extracted groundwater is expected to more than double, leading to an increase in the critical blocks. Low flow values in the rivers are predicted to decline compared to present levels...water quality and environmental flow conditions already critical will deteriorate further'. This does not augur well for the future. Some incentives premised on realities of that particular time become rather impossible and harmful to endure.

Another initiative undertaken by India can have positive concomitant impact on Bangladesh. In May 2015, the Indian government launched 'Namami Gange' project to 'conserve, clean and rejuvenate' the Ganga river to achieve cleanliness of Ganga by March 2019. But recent testing of water suggests that there is a significant rise in coliform bacteria and bio-chemical oxygen demand (BOD), important parameters to evaluate water quality.[26] This suggests that there is a missing link between promise and actual realization. The important point is that water planning and management cannot be effective in isolation, riverine countries have to engage each other in planning and execution for well-being of people.

CONCLUSION

The above analysis underlines that Himalaya is the powerhouse of water supply to Bangladesh and its co riparian countries. Some sort of understanding must be developed by the riverine neighbours to maintain the health of this fragile ecosystem; this would ensure the assured uninterrupted conveyance of water to the deltaic zone. Surface and groundwater are interconnected. Groundwater should be treated as reserved and policy should be in place to replenish this reserve as much as it is used. Maintaining health of the groundwater is very important in Bangladesh's climatic conditions. Apart from these interconnected issues, Bangladesh has to tailor agriculture policy in synergy with water availability; these initiatives are urgent with climatic change phenomena becoming apparent.

Bangladesh is rich with the presence of micro-water bodies like ponds, lakes. These need to be revived and activated. State cannot work as a monolith organization to provide one-stop solution to various concerns of water

---

[26] Singh, B. (2019, March 15). Ganga Water Quality Has Worsened in 3 Years: Study. *Times of India*.

management; it needs to have micro-interactive spots to test and filter the ideas at various levels like local government agencies, community-based institutions, management of ponds, water-bodies, watersheds, aquifers and river basins.

Water Issues in Bhutan: Internal Disputes and External Tensions

Rajesh Kharat and Aanehi Mundra

Bhutan has a very small territorial area compared to its neighbours. The pattern of the stretch of its geographical area of, approximately 47,000 km², makes it a compact country with maximum north-south distances of 170 km. and maximum east-west distances of 300 km. The unique geographical location and landlocked nature of Bhutan isolates her from the rest of the world. According to the Surveyor General of India, Bhutan is situated along the southern slopes of the great Himalayan range. The north of Bhutan is bounded by Tibetan plateau, highest plateau in the world, whereas on the south it is surrounded by the plains. In west, the strategic Chumbi valley of China, Sikkim and Darjeeling are present, while the eastern side is the Kameng district of Arunachal Pradesh in India.

Bhutan does not have any land or sea route to access other countries, other than by crossing India and China. Bhutan maintains its external and trade relations through Indian territory. The nearest seaport for Bhutan is Kolkata, which is 750 kms away from Phuntsholing. The mountainous

---

R. Kharat (✉) · A. Mundra
Centre for South Asian Studies, School of International Studies,
Jawaharlal Nehru University, New Delhi, India

R. Kharat
Equal Opportunity Office, Jawaharlal Nehru University, New Delhi, India

© The Author(s) 2020
A. Ranjan (ed.), *Water Issues in Himalayan South Asia*,
https://doi.org/10.1007/978-981-32-9614-5_3

terrain and thick forest are some physical features for Bhutan which make inaccessible to traverse the stretch, other geographical factors include climate conditions, heavy rainfall. This has impacted the construction of roads and railway connections. The only possible transport is by air, which is in Paro and not in Thimphu, the capital.

Apart from the physical accessibility, another feature of Bhutan is its sparse population. The geographical features of Bhutan have compelled it to have a limited population in any region, which results in the population nowhere exceeding more than 10,000 or 15,000. In this resource-rich region lies the small Himalayan Kingdom of Bhutan with a population of around 735,553.[1] About 57% of total population is involved in agricultural activities. The total urban population in 2017 was 274,316 persons and the rural population was 452,829.

The mountainous region and thick forest of Bhutan is also responsible for the isolation within Bhutan. Bhutan is not only isolated from the rest of the world but also isolated from within. As a result, the geographical features impact economic activities in the country. Although, the Bhutan's economy is mainly sustained by agriculture and animal husbandry, its production of food grains is insufficient due to the natural environment and non-availability of infrastructure like irrigation, seeds, new scientific methods. The mountainous terrain becomes a hurdle for smooth internal transportation and communication in Bhutan.

In the power sector, Bhutan has only two sources for its power and fuel resources that is, coal and hydroelectric energy. Since coal reserves are limited in quantity as well as quality, Bhutan imports almost all its coal from India. The other source of energy is hydroelectric power, which depends on rivers like the Torse, the Raidak and the Manas. However, only a very small portion of this enormous hydropower potential has been tapped, because of inaccessible areas and lack of capital investment and technical knowhow for development. The Government of India is helping Bhutan in this sector in various ways. Bhutan and India, has been signing various agreements and pacts since September 1961 in order to harness the hydro-electricity resources to purchase the ensuing power from Bhutan for its power-deficit states in north and north east India.

---

[1] National Statistical Bureau. (2018). *Statistical Year Book*. Retrieved from http://www. nsb.gov.bt/publica-tion/files/SYB_2018.pdf.

Bhutan transitioned to democratic system of governance in 2007, where King still holds a major place in the governance process. It has 20 Dzongkhags, the units for an efficient administration.

The GDP per capita of Bhutan is $2401. Bhutan is considered under the Least Developed Countries in the world since its inception in the list in 1971, after 2018 biennial review it is scheduled to graduate out of the list by 2023.[2] Bhutanese believe in Gross National Happiness as the measure of their growth, which ensures their developmental model is sustainable.

The Mountainous Kingdom of Bhutan is considered as the last Shangri-La. It is endowed with immense natural wealth of which 72% area is covered with forest and about 7.5% is covered by snow and glaciers.[3] The major water resources include: rivers, glaciers and groundwater. This water-rich nation has one of the highest per capita availability of 109,000 cubic metres every year, as compared to other south Asian countries. The significance of water emanates from the idea that it is the basic unit of life for sustaining the ecosystem. For Bhutan hydropower potential gives major economic benefits, while multiple rivers are responsible for manifestation of Bhutanese society. The water availability for Bhutan is fulfilled from rivers, glaciers and underground resources.

PRESENT WATER RESOURCES IN BHUTAN

Rivers

Rivers form the lifeline of civilizations, they are the cradles where life forms nourish. Bhutanese terrain of Himalayas is steep and gifted with broad drainage system. A river basin is the entire land area, including mountains and valleys, into which all waters—ice melt, snow, lakes, rainwater and groundwater— flow into, and merge with, a specific river that exits the area at one point. Rivers are mostly fed by rainfall, estimated 2–12% glacial melt and 2% snow melt. The combined outflow of the rivers is estimated

---

[2] Economic Analysis & Policy Division. *Least Developed Country Category: Bhutan Profile.* Retrieved from https://www.un.org/development/desa/dpad/least-developed-country-category-bhutan.html.

[3] National Environment Mission. (2016). *National Integrated Water Resources Management.* Retrieved from http://www.nec.gov.bt/nec1/wp-content/uploads/2016/03/Draft-Final-NIWRMP.pdf.

at 70,576 million m3, or 2238 m^3/s.[4] There are four major rivers systems in Bhutan:

1. Amochhu (or Torsa), is the 358 km long-smallest river system. It flows out of Tibet and enters western Bhutan via the Chumbi Valley. It passes the India-Bhutan border town of Phuntsholing and then flows into India. The Amochhu basin covers the districts of Haa, Samtse, and Chhukha. Its total catchment area is 2298 km^2, which is about 6% of the country's total land area.[5]
2. Wangchhu (or Raidak), is 370-kilometer-long river. Its major tributaries include, Pachhu, Tachhu, Hachhu, Thimchhu and Wongchhu, which mostly rise in Tibet. The Wang Chhu in its course drains the populated civilization areas of Ha, Paro, and Thimphu valleys. The Wangchhu enters West Bengal Duars as the Raigye Chhu after flowing southeasterly through west-central Bhutan. Spread over 11% of the country's total land area, the Wangchhu basin has an area of 4596 km^2.
3. Punatsangchhu (or Sunkosh) rises in northwestern Bhutan and the 320-kilometer-long river flows through Punakha, the ancient Bhutanese capital before entering into West Bengal in India. Its tributaries include, Mochhu, Phochhu, Tangchhu, Harachhu, Dagachhu, Basochhu and Dan gchhu. The largest in Bhutan, the Punatsangchhu basin covers 9645 km^2, which represents 25% of the country's total land area.[6]
4. Manas (or Drangmechhu) is a major river system in India and Bhutan. Its tributaries include, Mangdechhu, Kurichhu, Chamkarchhu,

---

[4]National Environment Mission. (2016). *National Integrated Water Resources Management*. Retrieved from http://www.nec.gov.bt/nec1/wp-content/uploads/2016/03/Draft-Final-NIWRMP.pdf.

[5]Asian Development Bank. (2016). *Water-Securing Bhutan's Future*. Retrieved from https://www.adb.org/sites/default/files/publication/190540/water-bhutan-future.pdf.

[6]Asian Development Bank. (2016). *Water-Securing Bhutan's Future*. Retrieved from https://www.adb.org/sites/default/files/publication/190540/water-bhutan-future.pdf.

Drangmechhu and Kholongchhu. It flows from India towards Eastern Bhutan and drains the Tongsa and Bumthang valleys.[7] It covers 8457 km^2, which represents 22% of Bhutan's total land area.[8]

Apart from the main river system, other smaller rivers originate and flow in the Middle Hills, namely the Jaldhaka, Mao, Badnadi and the Dhansiri. These rain-fed rivers are big and turbulent in the monsoon, whilst run almost dry during the winter.[9]

All the river systems originate within the country, except three rivers, viz. Amo Chhu, Gongri and Kuri Chhu, all of which originate in the southern part of the Tibetan Plateau.[10] The Southern flow of all river systems fulfills the need of the country, by touching the life of the entire nation. Except for a small river in the extreme north, all the rivers enter India, where they join the Brahmaputra River system. The river basins are oriented north-south and are, from west to east.

Most of the rivers are deeply incised into the steep geography of Himalaya which makes the possibilities for run-of-the-river irrigation projects limited.

Glaciers

Bhutan is a Himalayan state with the presence of snow-covered peaks and glaciers. According to a survey mentioned in Bhutan's National Integrated Water Resources Management Plan, there are 885 Clean Glaciers and 50 Debris Covered Glaciers in 2010 covering an area of 642 and 16.1 km^2

---

[7]The flow of river is mentioned in country studies of US Library of Congress. Retrieved from http://countrystudies.us/bhutan/16.htm.

[8]Asian Development Bank. (2016). *Water-Securing Bhutan's Future*. Retrieved from https://www.adb.org/sites/default/files/publication/190540/water-bhutan-future.pdf.

[9]Dhakal, D. N. S. (1990). Hydropower in Bhutan: A Long-Term Development Perspective. *Mountain Research and Development*, X(IV), 291–300.

[10]Kusters, K., & Wangdi, N. The Costs of Adaptation: Changes in Water Availability and Farmers' Responses in Punakha District, Bhutan. *International Journal Global of Warming*, V(IV), 387–399.

56 R. KHARAT AND A. MUNDRA

respectively. The height of glaciers ranges from 4050 to 7230 m above the sea level.[11]

As per Kathmandu-based International Centre for Integrated Mountain Development, there are even larger numbers of Glacier Lakes, 2674 with an area of 106.87 km^2, out of which 24 are Potential Glacial Lake Outburst Floods (GLOF). This leads to disaster in the nearby areas as large quantity of water floods limited areas.

Glaciers are source of rivers and hence the lifelines of the society. They are also an indicator of climate change as any global change in temperature is noticed on them. Increased rate of melting of glaciers leads to overflowing of rivers, often leading into floods. This direct visual evidence of glaciers as it alters their thickness, area and tail end; features which are result of the change in the atmosphere around them.

Groundwater

There is not enough data to analyse the groundwater resources in Bhutan. However the knowledge of availability of groundwater may solve the issues of scarce regions. In other parts of the world, the groundwater is the source of some major rivers. The geography of Bhutanese Himalayas consists of steep terrain and deeply incised valleys. The present available information maintains that there is no real groundwater aquifer, although there should be subsurface flow through recent deposits. Some valleys are comparatively wider and flatter for example Paro, Punakha, Thimphu. The plains of Phuentsholing, which borders India's West Bengal may have groundwater reserves that could be exploited. There may be extraction of groundwater resources by individuals at present, but government is not using groundwater as long as the sustainability of its use has not been assessed.

Numerous reservoirs in main rivers or main tributaries have been built for hydropower generation. These reservoirs are also useful for the purpose of flood attenuation. It is difficult to create many reservoirs in Bhutan's terrain because of geologically seismic topography. There are also plans to build reservoir-type hydroelectricity plants in Amochhu, Bunakha and

---

[11] National Environment Mission. (2016). *National Integrated Water Resources Management*. Retrieved from http://www.nec.gov.bt/nec1/wp-content/uploads/2016/03/Draft-Final-NIWRMP.pdf.

Sankosh.[12] These reservoirs may be sufficient to address the water needs of neighbouring areas.

INTERNAL WATER ISSUES IN BHUTAN: LOCAL ISSUES

Although Bhutan is endowed with great natural wealth, the country deals with various internal water issues, which also include water shortages in certain seasons. Mostly internal water issues can be bifurcated as natural and manmade.

Natural Issues

Bhutan is a geologically sensitive and seismic zone. It means it is hugely prone by risks of geographical disasters, which mainly include earthquakes and GLOF. It happens due to the tough terrain of mountainous Himalayan region.

During any construction, landslides are accentuated with unfavourable monsoon and loose soil structure of the terrain, which increases possibility of soil erosion.[13]

GLOFs are caused by excessive melt water bearing down on moraine dams. These GLOF events have effects across the boundary.

Other issues which impact the water balance include flash floods, sediment run off, drying of springs. The natural issues aren't under human control nor we can stop them by our limited resources and understanding. Also, we need to understand that their impact is exacerbated by the events of Climate Change, which has increased recently due to anthropogenic factors in recent times.

Man Made Issues

These issues arise due to Human interference in the natural working of things, availability of clean drinking water, impact of construction of dam

---

[12]National Environment Mission. (2016). *National Integrated Water Resources Management*. Retrieved from http://www.nec.gov.bt/nec1/wp-content/uploads/2016/03/Draft-Final-NIWRMP.pdf.

[13]Vasudha Foundation. (2016). *A Study of the India-Bhutan Energy Cooperation Agreements and the Implementation of Hydropower Projects in Bhutan*. Retrieved from http://www.vasudha-foundation.org/wp-content/uploads/Final-Bhutan-Report_30th-Mar-2016.pdf.

and storage of water and impact on agriculture and other allied small scale industries due to water. This gives rise to bother linked issues, like waste water treatment, internal disputes, etc.

Like any other country on path of development, rising industrialization has also led to deteriorating quality of water in some pockets of Thimpu.[14] Although population is less, there are only two wastewater collection and treatment projects in the cities of Thimphu and Phuntsholing.

The potential of water has brought the maximum economic prosperity with the growth of Hydropower sector. The concerning issues in the hydropower sector include: prediction future flows; management of hydropower systems for future flows; sedimentation of reservoir; floods, including flash floods and GLOFs; increase in glacier retreat and less snow cover; and erratic rainfall patterns. The impact of natural factors also impacts the manmade structures.

The harnessing of Hydropower potential has led to internal tensions and other related issues within Bhutan and its lower riparian areas, which will be discussed in detail in a later section.

Hydropower Project and the Economy of Bhutan and Power Debt to India

As Bhutan is an under-developed country, its planners as well as those in India realized the pressing need for electric power in Bhutan to boost its economic growth.

In order to achieve that objective, in September 1961, Bhutan and India signed an agreement for a joint HydroElectric Project on the Jaldhaka river.[15] This river runs along the south-western, Indo-Bhutanese border for about twelve miles. Bhutan received its first electricity in the spring of 1968. But most of the benefit from the project had gone to West Bengal. However, Bhutan received Rs. 2.5 million, as royalty from West Bengal, for use of Bhutan's river water. Over and above, Bhutan was able to get 250 kilowatts of electricity, without cost.[16]

---

[14]Giri, N., & Singh, O. P. Urban Growth and Water Quality in Thimphu. *Journal of Urban and Environmental Engineering*, *VII*(I), 82–95.

[15]Belfiglio, V. J. (1972, August). India's Economic and Political Relations with Bhutan. *Asian Survey*, *12*(8), 678.

[16]Seymour, S. (1978, March). Strategic Development in Bhutan. *Strategic Digest*, *III*(3), 56.

After the successful implementation of this project there were two more hydroelectric projects which had been completed in Bhutan with Indian assistance. The fourth project began in 1974 and was administered by the Chukha Project Authority. This was the Chukha Hydro Power Project, which was set up according to an agreement between the Governments of India and Bhutan. This project located at a place named Chukha, about fifty-five miles away from the capital Thimpu, along with the Raidek river. This project was entirely financed by the Government of India. 60% of its fund was given as a grant, and the remaining 40% of fund was a loan. It is the largest power project in Bhutan with a capacity to generate 336 megawatts (MW). India financed Rs. 2040 million for this project.[17] The Government of India also financed other hydropower projects like the Kurichu Hydroelectric Project (46 MW), the Tangsibi Hydro-electric Project (45 MW), the Basochu Hydroelectric Project (46 MW), the Bunakha Reservoir Project (120 MW), Chukha II (1000 MW) and Chukha III (600 MW).

These hydropower projects will result in a reduction of the cost of electricity supplied to the consumers. They also provide a stepping-stone to the development of the infrastructure in Bhutan's economy, besides being a major source of revenue through the sale of power to India, given the latter's growing need of power.

The Chukha HydroPower Corporation (CHPC) is the biggest and the most successful of many projects in Bhutan which were implemented with India's technical and financial assistance. It was inaugurated on 21 October 1988 by Shri R. Venkatraman, the then President of India. On this occasion, the King of Bhutan appreciated the Indian financial and technical assistance and mentioned that this project was a symbol of Indo-Bhutan friendship and cooperation. King Jigme Singye Wangchuk said, 'This project could not have been undertaken if India, our very close friend and neighbour had, not come forward so generously with men, money and materials'.[18]

The cost of this project was estimated at that time to be around Nu. 2460 million and it is the biggest source of generating project for Bhutan. This project exports 95% of the power it generates to India through a substation in Birpura in West Bengal. This project is located on the Raidek river. On

---

[17] *Bhutan: Development Planning in a Unique Environment* (pp. 45–47). World Bank Report, 1989 (Washington: World Bank).

[18] Bhutan: An Outline of Its History, Political Institutions and Organisations of Government. *The Hindu* (Madras), November 11, 1992.

the same river another project was taken up, known as the Bunakha Reservoir Scheme, with Indian assistance. In addition to this, Bhutan signed a memorandum of understanding with India in 4 January 1993 for the development of the Sankosh Multi-purpose Project. This project is one of the largest multi-purpose projects in South Asia and generates 1525 MW of power. Besides, generating power, it irrigates 5,00,000 hectares of land in West Bengal.[19] Thus, the project provides immense developmental benefit to the people of Bhutan as well as India, in terms of power, water and revenue.

At this time, the King of Bhutan described Indo-Bhutan economic cooperation as based on complete trust and friendship between the two. Since then, India is paying the higher rate of tariff for Bhutan's surplus electricity provided extra revenue for the country's development. Earlier, on 27 September 1995, a power agreement had been signed between the countries under which the Government of India had agreed to fund the construction of the 45 MW Kurichu Hydroelectric project.[20]

In this way, these power projects with Indian assistance for support, proves to be a major source of income for Bhutan and to upgrade the standard of living of the people. It is important to note that these power projects are treated as symbols of Indo-Bhutan friendship.

India has been a supporter of Bhutan at international level and in domestic matters. India is the lower riparian state for the rivers flowing through Bhutan. The transboundary cooperation between the two countries has led to creation of a Joint Group of Experts (JGE) which discusses/assess the issues between India and Bhutan, which mainly include the recurring floods and erosions in the bordering foothills of Bhutan and Indian plains. The Experts Group gives regular inputs to both governments. The last meeting of JGE was held in April 2017 in Thimphu, Bhutan.[21] This mutually beneficial relationship provides clean electricity to India, along with generating export revenue for Bhutan.

---

[19] King of Bhutan, His Majesty Jigme Singye Wangchuck Holds Talks with Indian leaders, Foreign Affairs Record (MEA Publication, GOI, New Delhi). Vol XXXIX Vol I January 9, 1993.

[20] Cooperation With Neighbouring Countries in Hydro Power, Annual Report 1996–97, (Ministry of Power, GOI, 1997) p. 23.

[21] Ministry of External Affairs of India. (2017). *India-Bhutan Relations.* Retrieved from http://www.mea.gov.in/Portal/ForeignRelation/Bhutan_September_2017_en.pdf.

Hydropower sector is the potential earner for the Bhutan's economy such that the export of hydropower energy generates around 40% of the Bhutan's revenue. Hydropower sector's contribution to the country's gross domestic product (GDP) is about 25%.[22] Druk Green Power Corporation, operates and maintains hydropower assets of Bhutan and is the highest taxpayer of the country.[23]

The potential of Hydro energy can be analysed by the stark increase in the GDP with the commissioning of the fifth hydropower project in 2005. The leap was huge that in 2007 the Bhutanese economy became the second fastest growing economy in the world. During the period, Bhutan reached an annual growth rate of 22.4%.[24] The rugged mountainous terrain and swift flowing rivers have made Bhutan a natural haven for hydropower. At present, the hydropower potential of Large Hydropower plant is 30,000 MW and the total installed capacity is 1606 MW (5.35%).[25] But lately, hydropower has been declining over the years, from 44.6% in 2001 to 20% in 2013.[26]

The initiation of Bhutan's tryst with hydropower began with Indo-Bhutan hydropower cooperation in 1961. It began with the Jaldhaka project situated on the Indian side of Indo-Bhutan border in West Bengal, where power was exported to southern Bhutan by India. In 1987, first hydropower project 'Chukka' hydropower with potential of 336 MW was introduced in Bhutan by India. As per India–Bhutan Agreement on Cooperation in Hydropower in 2006 and the Protocol to the 2006 agreement signed in March, 2009, Bhutan will increase its generating capacity in 2013

---

[22] Ranjan, A. (2018, October 17). *India-Bhutan Hydropower Projects: Cooperation and Concerns* (ISAS Working Paper No. 309). National University of Singapore.

[23] Royal Bhutanese Embassy in India. (2016). *Bhutan-India Hydropower Relations.* Retrieved from https://www.mfa.gov.bt/rbedelhi/?page_id=28.

[24] Vasudha Foundation. (2016). *A Study of the India-Bhutan Energy Cooperation Agreements and the implementation of Hydropower Projects in Bhutan.* Retrieved from http://www.vasudha-foundation.org/wp-content/uploads/Final-Bhutan-Report_30th-Mar-2016.pdf.

[25] Energy Status in Bhutan. (2016). Retrieved from https://unstats.un.org/unsd/energy/meetings/2016iwc/18status.pdf.

[26] Yashwant, S. (2018, 4 September). *How Villagers in Bhutan and India Came Together to Resolve a Water-Sharing Tussle.* Scroll.in. Retrieved from https://scroll.in/article/892235/how-villagers-in-bhutan-and-india-came-together-to-resolve-a-water-sharing-tussle.

to six times by 2020. This will increase the targets to produce an additional 10,000 MW.[27]

As per brief by Bhutanese government, many projects of thousands of MW potential are under construction.

In April 2014, India and Bhutan signed the 'Framework Inter-Governmental Agreement' for development of Joint Venture Hydropower Projects through the Public Sector Undertakings. Under this Inter-Governmental agreement framework, cooperation for implementation of other power projects was taken up.[28]

While ascertaining Bhutan's need to grow, Biswas in his article 'Cooperation or conflict in transboundary water management: Case study of South Asia' has shown 'Water' as the protagonist which can bring the Bhutanese society in the realm of development and improving the lifestyle of people. On the other hand, it is beneficial for Bhutan to have capital and technological expertise from India. Also, the water potential cannot be distributed within the country itself as the population is small and segregated, which makes it difficult to have a dedicated absorbing capacity. But some newspaper reports mention that after India declared it power surplus, price of Bhutanese power has seen a drop.

'Bhutan's entire economic hopes have been pinned on selling hydropower to India', Tenzing Lamsang, the editor of *The Bhutanese* said in the commentary.[29] The issue of hydropower debt is huge with Bhutanese politics such that, Mr. Tobgay had promised to curb the national debt, owed mostly to India for hydropower loans in his last campaign.[30] It has always been but this time it has tilted towards India to such extent Bhutan felt that they have been exploited by Government of India.

Though the relations between the two nations are cordial, also reiterated by the brief of Bhutan's Foreign ministry. The relations are strained as the

---

[27] Royal Bhutanese Embassy in India. (2016). *Bhutan-India Hydropower Relations.* Retrieved from https://www.mfa.gov.bt/rbedelhi/?page_id=28.

[28] Royal Bhutanese Embassy in India. (2016). *Bhutan-India Hydropower Relations.* Retrieved from https://www.mfa.gov.bt/rbedelhi/?page_id=28.

[29] Zhen, L. (2018, 20 July). Is Bhutan Drawing Closer to China, and What Can India Do About It? *South China Morning Post.* Retrieved from https://www.scmp.com/news/china/diplomacy-defence/article/2108804/bhutan-drawing-closer-china-and-what-can-india-do-about.

[30] Haider, S. (2018, 20 August). Sovereignty and Sensitivity: On India-Bhutan Relations. *The Hindu.* Retrieved from https://www.thehindu.com/opinion/op-ed/sovereignty-and-sensitivity/article24731900.ece.

trade deficit in favour of India reached US\$150 million of US\$516 million of Bilateral trade between the two nations in 2015.

Apart from the queries on hydropower issues in Bhutan, a Public Debt Policy[31] was adopted in 2016 to contain public debt. The debt is considered as an indicator of curtailing the openness of hydropower relations of India and Bhutan, as hydropower has huge stake in Bhutan's GDP and its bilateral relations with India. This policy is an indication of the new path that relations may take between these two countries.[32]

The hydropower debt has remained a serious issue till date. There has been questioning in National Assembly on the rising debt of the country due to power projects. In the first question hour session on 17 November 2017, Prime Minister (PM) was asked about the rising national debt on Kholongchhu hydroelectric project. The PM responded by quoting the 'Inter-Governmental Framework Agreement' with the Indian government would be pursued for a solution.[33]

The gap in understanding between the two nations was also realized by Indian politics, so as to reduce it in 2014, Prime Minister Narendra Modi, as he came to power in India made Bhutan his maiden visit abroad.[34]

WHICH INSTITUTIONS ARE RESPONSIBLE FOR WATER RESOURCES IN BHUTAN?

Water is a flowing precious resource which surpasses different facets of human society. The present system of governance in any country can't limit the maintenance of water resources to one institution. Mountains are the sources of rivers and have been revered as the home to a large proportion of the world's population throughout history. However, the country's mountain ecosystem is becoming more vulnerable to the increasing threats from climate change. In Bhutan major water potential is harnessed by the

---

[31] Ministry of Finance. Public Debt Policy 2016. Retrieved from http://www.mof.gov.bt/wp content/uploads/2014/07/PublicDebtPolicy2016.pdf.

[32] Ibid.

[33] Wangmo, T. (2017, 19 November). Hydropower at the Heart of NA's Question Hour. *The Journalist*. Retrieved from http://www.bhutanjournalist.com/journalist/hydropower-at-the-heart-of-nas-question-hour/.

[34] Stobdan, P. (2017, July 14). India's Real Problem Lies in Its Bhutan Policy, Not the Border. *The Wire*. Retrieved from https://thewire.in/diplomacy/india-china-doklam-real-problem-bhutan.

hydropower plants and the Department of Power has been given responsibility for hydrological and meteorological data collection. The water use for agriculture is dealt with Irrigation Agency under Ministry of Agriculture, which comprises of irrigation offices at three levels. Nevertheless, Bhutan is fortunate and blessed with rich water resources and the highest per capita[35] of water availability is 109,000 cubic meters.[36] Despite, there is always a concern of water shortage in most parts of the country. Fresh water is distributed throughout the country from snow, glaciers and lakes to springs, streams and rivers. The Royal Government of Bhutan is keen on providing clean, safe drinking water to all citizens in towns and villages.

Bhutanese government has an Environmental agency dedicatedly working on environmental issues in Bhutan. The National Environment Commission (NEC) works for effective coordination of national-level planning and development for all natural resources including water resources, formulation of water policy and the necessary legislation. As recent in 2016, NEC launched National Integrated Water Resource Management Plan. National Environment Commission Secretariat (NECS) works in assisting the provisions in the Water Act. A Water Resource Coordination Division was set up in May 2010.

The Bhutan Water Policy was brought in 2003 which was an important legislation on water among others. Apart from focusing on much needed issues, water user interests and priorities, principles for water resources development and management, it also focuses on institutional development for water resources management.

Bhutan Water Policy 2007 gave a vision statement on water. While maintaining that water is the most important natural, economic and life-sustaining resource, it mentioned, 'we must ensure that it is available in abundance to meet the increasing demands'.

The policy also focusses on providing the adequate amount and accessibility of safe and affordable water to the present and future generations.

With increasing consumptive demand, competition for water is emerging, which may ramble the prosperity and happiness of the nation. For Bhutan, it is a major contributor in the policymaking structure. The Gross

---

[35]With 109,000 cubic metre of per capita freshwater, Bhutan has been ranked 6th among 200 countries which is highest in the region.

[36]National Water Symposium Brings Experts Together. *The Bhutan Times*, May 14, 2017.

National Happiness Commission is the apex body for setting the development priorities and plans for the country. Planning officers across the country support the Secretariat with planning and monitoring progress of local and sectoral plans.

HYDRO PROJECT AND ITS SOCIAL IMPACTS ON BHUTAN

Bhutanese often refer to their water resources as 'white gold'.[37] It is worth the title as it has been crucial in bringing prosperity to the small nation which traversed on the developmental path late. Water has given boost to economy along with impacting the society in every form of existence. As hydropower has brought electricity export, it also brought local and internal issues: less involvement of Bhutanese private sector, more Indians migrate to work in the projects than locals in Bhutan and reduced economic cooperation with increased interest rates is leading rise in Bhutan's debts.[38]

While all the positives of power projects have created a wave of appreciation for India's work in Bhutan's hydroelectric potential. As per newspaper reports, some Bhutanese locals also protested that such projects can create issues related to water storage for Bhutan. Environmentally it is unsustainable to store large water in seismic regions of Bhutan, wherein some projects created by India may be so, and Bhutanese may not be able to object. Also Bhutan has internal water shortage, so it may want to create reservoirs for fulfilling the need of its people, which may not be favourable for India as it is a lower riparian state. Apart from that there are fears of frequent flooding, erosion and destruction. These tensions have been further accentuated by recent erratic weather patterns.[39]

The weather patterns have been troublesome as it leads to seasonal variations of river flows. The economic and environmental implications of

---

[37] Asian Development Bank. (2016). *Water-Securing Bhutan's Future*. Retrieved from https://www.adb.org/sites/default/files/publication/190540/water-bhutan-future.pdf.

[38] Mahajan, S. (March 6, 2018). *India-Bhutan Relations: Past, Present and Future*. South Asia Program at Hudson Institute. Retrieved from http://www.southasiaathudson.org/blog/2018/3/6/india-bhutan-relations-past-present-and-future.

[39] Khan, M. N. (2017, 18 January). India's Water Issues with Bangladesh, Nepal, and Bhutan. *The London Post* Retrieved from https://thelondonpost.net/indias-water-issues-with-bangladesh-nepal-and-bhutan/.

hydropower owing to these reasons are well known. Primarily, 'run-off-the-river' hydropower plants are constructed. They are concentrated in the middle to lower parts of the basins.

These are non-reservoir based and non-consumptive, it leads to fluctuation in power generation, especially during the winter when river water is inadequate for the optimal utilization of the plants. This fluctuates the economic stability of an entire nation.

The Government of Bhutan allocates at least 1% of annual hydroelectricity royalties to the Ministry of Agriculture and Forests for sustainable agriculture and upstream catchment protection.[40]

Unemployment

Bhutan has a small population, but it also has limited opportunities. Its major economy is sustained by hydropower followed by urban infrastructure development. However, the favourable tourism potential is exploited sustainably which is beneficial for environment and caters to the Gross National Happiness achievement.

Bhutan explores its hydropower by creation of its run-of-river projects; labour issue is involved in the construction of these dams. Some reports mention that major population which work on the projects, migrate from India, as also India is a major player in harnessing of the hydro potential. This movement from India is due to population-opportunity imbalance in India, they serve as cheap labour for Bhutanese economy. Mass movement of foreigners increases the demographic pressure on adjacent natural resources. (NCD 2009) An article, 'Bhutan's Happiness Saga' in South Asian journal[41] mentioned that this has also been a reason for Bhutan's low rank of 97 out of 156 countries in United Nations World Happiness Report (WHR). The out-migration of Bhutanese youth aggravates one of Bhutan's pressing problems—youth unemployment. Bhutan's rankings trail behind countries, many of which unrest frequently disrupt daily life.

---

[40] Asian Development Bank. (2016). *Water-Securing Bhutan's Future*. Retrieved from https://www.adb.org/sites/default/files/publication/190540/water-bhutan-future.pdf.

[41] Rashed Nabi, Bhutan's Happiness Saga, October 2, 2018. Retrieved from http://southasiajournal.net/bhutans-happiness-saga/.

Environmental Issues

Environment is the primary focus of Bhutan's GNH philosophy. Their tourism requires every foreign tourist to spend $250 per day for their stay in the country. These laws mention the presence of a serious concern for environment. The collaboration for hydropower would bring the beneficial fruits along with the necessary concern for the exploitation of resource. This leads to an understanding of exploring resources on more sustainable manner. The Indian government's Central Electricity Authority, an agency for power sector development, identified about 76 locations in Bhutan for hydropower projects.[42] For materializing this plan of 23,760 MW electricity, dams would have to be built over every river.

This may have raised eyebrows in Bhutan as maximizing its full hydropower potential may not maximize its sustainability. However, the Environment Impact Assessment reports of neither Bhutan nor India have been on their websites.[43] While other issues of the sediment and biota flows, the health of riverine ecosystem and its impact on downstream population remains.

Drinking Water Problem

While conflicts among people are considered an indicator to gauge the severity of the water problem, other issues are severe too. An article in the third pole mentioned, 'Around 700 acres of land in the area has been left fallow and some farmers have migrated to towns because of the lack of water.

There are localized and seasonal water shortages for drinking and agriculture in Bhutan. Only 78% of the country's population has access to safe drinking water and about 12.5% of the arable land is irrigated. As per National Statistics Bureau data (2017) Population Access to Improved Water Sources is 98.6%. The figure is decent, but the quality of water in Bhutan is deteriorating can be ascertained by the launch of government's drinking water policy in 2016. In 2014, Bhutan's health ministry conducted an inventory of rural households, which brought out results that

---

[42] South Asia Team. (2015, 16 June). Bhutan Rivers Watch. *International Rivers*. Retrieved from https://www.internationalrivers.org/resources/9059.

[43] South Asia Team. (2015, 16 June). Bhutan Rivers Watch. *International Rivers*. Retrieved from https://www.internationalrivers.org/resources/9059.

Fig. 1 Riverside in Paro. Aanehi Mundra. April 2016

17% of the total number of households faced drinking water problems in Bhutan. The cost to deliver water in the physical terrain of Bhutan is tough, lack of investments in such terrain make it even worse.[44] The government reported that in the 12th Plan, all projects for drinking water would be taken up as flagship programmes (Fig. 1).

Problem of Accessibility of Water Resources

The accessibility of water for the settlements on the slopes is an issue, while the major rivers flow nearby at the valley bottoms. People are inflicted with such shortage seems strange, as per capita water availability of Bhutan is one

---

[44]Gyelmo, D. (2016, 21 April). *Bhutan Struggles with Local Water Shortages*. Thethirdpole.net. Retrieved from https://www.thethirdpole.net/en/2016/04/21/bhutan-struggles-with-local-water-shortages/.

of the highest in the region, in a situation where both Bhutan's populated neighbours India and China are in stress.

Fifty-six rivers traverse the geography from Bhutan to India. An article recently published in scroll mentioned that, while Bhutan hills are lush green, Indian side of the border comprises of dry patches of denuded forests and vast tracts of dry plains. There are many such initiatives too where several locals come together to participate in creating small check dams and consuming water. The Bodo tribe created traditional diversion based irrigation system called the Dongo or Jamfwi system; it was created at the borders by about 500 farmers over the Saralbhanga River (also called Swrmanga) which flows from Sarphang district of Bhutan to Assam in India.[45]

The issue of accessibility of water has become a part of election manifesto, as PDP mentioned that if they were re-elected, all homes would be provided water 24 × 7. It is an irony that a water-rich country's political manifesto includes provision for continuous water to its people.[46]

Understanding the community collaborations may be more useful if right measures are taken. When we have success stories as mentioned, diarrhoea is still one of the country's top ten diseases.[47] In a cold country, tropical diseases like dengue and malaria are also prevalent, which indicates the poor sanitation and absence of clean water.[48] More challenges tend to rise in future, as with a population growth rate of 1.3% per annum, Bhutan's population will double in the next fifty years. Almost fifty per cent of this population is expected to live in urban centres. This rising populace may create crunch for the limited resources and delivery of services.[49]

Besides, Bhutan's major economic sectors are already witnessing, the shortage to adequate and consistent water supply. For instance, drying up of streams in some parts, and shortages in towns. This water stress has taken

[45]Yashwant, S. (2018, 4 September). *How Villagers in Bhutan and India Came Together to Resolve a Water-Sharing Tussle.* Scroll.in. Retrieved from https://scroll.in/article/892235/how-villagers-in-bhutan-and-india-came-together-to-resolve-a-water-sharing-tussle.

[46]Kuensel. (2018, 21 June). The Water Crisis. *Kuensel.* Retrieved from http://www.kuenselonline.com/the-water-crisis/.

[47]*Owning the Water Supply: Bhutan Water Quality Project.* Retrieved from http://www.searo.who.int/entity/partnerships/bhutan.pdf?ua=1.

[48]Jamtsho, T. (2010). Urbanization and Water Sanitation and Hygiene in Bhutan. *Regional Health Forum, 14*(1).

[49]National Environment Commission—Bhutan Water Policy. (2007). Retrieved from http://extwprlegs1.fao.org/docs/pdf/bhu167545.pdf.

in farming communities to reconsider their dependence on agriculture. Bhutan's river systems are heavily dependent upon monsoon and these rivers are already run low in the dry season. With the rising impact of climate change the rivers could display even more extreme flows in the years ahead, which could be further accentuated with increasing urbanization which is already affecting water quality.[50]

Internal Water Disputes

There are reports of rising discontent among people because of water unavailability. The internal water disputes have been severe, for example, in some communities of Samdrup Jongkhar District, like Orong, water shortage became the biggest problem as per reports in 2014. This problem impacted hygiene, drinking water availability, as well as irrigation and crop production. The impact on society has penetrated the Geog Administration such that it recorded 10 disputes related to water in a year.[51]

Institutional Fallout

Water is a resource which is catered to by numerous agencies. This creates a fallout in the coordination for its institutions have weak functional linkages. These gaps are realized at policy, planning and programming levels. All the sub-sectors work independently of each other, which lead to fragmented data and duplication of efforts. This can be seen in water management as urban water supply is under Ministry of Works and Human Settlements and rural water supply is with the Ministry of Health. This absence of a coherence has given space to potential conflicts.[52]

Apart from regulation, the policy guidelines by Bhutanese government are given for the quality of water too. This deteriorating quality of raw water sources in Bhutan is recognized as already a range of policies and regulations are in place since 2003 such as Water Resources Management

---

[50] Bhutan—Water Risk Scenarios and Opportunities for Resilient Development. (2016). Volume 1 and 2 (WWF Report Living Himalayas), Bhutan.

[51] Ministry of Agriculture and Forests of Bhutan. (2014). *Orong Farmers Solve Water Shortage*. Retrieved from http://www.moaf.gov.bt/orong-farmers-solve-water-shortage/.

[52] National Environment Commission. *Bhutan Water Vision 2025 and Bhutan Water Policy*. Retrieved from http://www.nec.gov.bt/nec1/wp-content/uploads/2014/04/Bhutan-Water-Policy-Eng.pdf.

plan, 2003. Other schemes inlcude the Rural Water Supply Scheme policy, the Water Act of Bhutan, 2011, the Water Regulation, 2014, the Environmental standard, 2010, and the watershed management guidelines that have been passed to ensure the sustainability and to protect the quality of drinking water sources.[53] So many schemes have been initiated over the years specifically rendering to the water issue, while the paradox is still in 2016, a Drinking water policy had been launched in Bhutan. Even a more water strained country like India does not have a specific drinking water policy.

International Waters: Transboundary Water Collaboration

For water sharing procedures, Bhutan's Water Policy mentions, 'Transboundary water issues shall be dealt in accordance with International laws and Conventions to which Bhutan is a signatory. and cooperation in information sharing and exchange'.[54]

Unauthorized transboundary water abstractions are potential issue of contention in transboundary water sharing. The National Integrated Water Resources Management Plan of 2016 has mentioned the reports of Bhutan's water resources being 'informally' tapped by people from across the border for personal usage.

Some reports of these abstractions mention that the pipes extended well over one km into Bhutan. Some were going-on since long and payments were made. This will impact the water availability in the local communities as it will create stress without payments made.[55] These issues have not come out in open and clear, but their mention in government documents may lead to rising problems with neighbouring state.

The Indian states of Assam and West Bengal share a border of 267 and 183 km respectively with Bhutan. The upstream 600 MW Kholongchhu HEP on the Manas river is a concern for ecosystems in the downstream

---

[53] National Environment Commission. *Bhutan Drinking Water Quality Standard, 2016*. Retrieved from http://www.nec.gov.bt/nec1/wp-content/uploads/2016/04/BDWQS-final.pdf.

[54] National Environment Commission. *Bhutan Water Vision 2025 and Bhutan Water Policy*. Retrieved from http://www.nec.gov.bt/nec1/wp-content/uploads/2014/04/Bhutan-Water-Policy-Eng.pdf.

[55] National Integrated Water Resources Management. (2016). Retrieved from http://www.nec.gov.bt/nec1/wp-content/uploads/2016/03/Draft-Final-NIWRMP.pdf.

72 R. KHARAT AND A. MUNDRA

in India. There are other upstream reservoir dam projects, located in the Manas and Sankosh river basin, which will have sociological and ecological impacts down-stream.[56] An article mentioned in Down to Earth (2016), focussed on the unannounced release of water from a Bhutanese dam which led to flash flooding. The Kurichu dam which is known for releasing water is situated upstream the River Kurichu, a principal tributary of the Manas.[57] The power plant is managed by Druk Green Power Corporation, which led to destruction in Lakhimpur districts in Assam, along with flooding of Manas forest area. There is an agreement between India and Bhutan to share information on the release of water from the Kurichhu project, but there are allegations against Bhutanese authorities of not fulfilling the needs of the agreement.[58] Although Bhutanese authorities have refuted these allegations mentioning that it is a small run-of-river project of 60 MW and they have always shared information of water release with Indian embassy, which may not have been shared to the Assam government.

The difference in the views of the transboundary neighbours needs to be dealt with proper dialogue channel and measures under the water policies.

How Do Locals Perceive Water and Deal with the Issues Linked?

There are nine basic values, as per Social ecologist Stephen Kellert, which are commonly applied to natural places and wild species. These include material benefits, source of knowledge, aesthetic beauty, companionship and relationship, mastery through challenge, morality/spirituality, natural-ism and wonder, fear and symbolism.[59] All these attributes logically apply to Bhutan's efforts to preserve and protect the natural resources without disturbing the serene nature and bio-diversity of Bhutan. Moreover, Bhutan

---

[56] South Asia Team. (2015, 16 June). Bhutan Rivers Watch. *International Rivers*. Retrieved from https://www.internationalrivers.org/resources/9059.

[57] Down To Earth. (2016, 18 October). Floods Affect over 30,000 in Assam. *Down To Earth*. Retrieved from https://www.downtoearth.org.in/news/natural-disasters/floods-affect-over-30-000-in-assam-56023.

[58] Ahmad, O. (2016, 18 October). *Massive Flood on Bhutan-India Border Triggers Blame Game*. Scroll in. Retrieved from https://scroll.in/article/819153/massive-flood-on-bhutan-india-border-triggers-blame-game.

[59] Kellert, S. R. (1996). *The Value for Life*. Washington, DC: Island Press.

government takes utmost efforts to maintain 'spiritual relationship' with the nature.[60]

The existence of Bhutanese society is manifested with religion and is reflected in worshipping of water. In a paper by Ugyen Wangchuck Institute for Conservation and Environment (UWCIERA) on locals suffering with the issues of water availability, 88% of the respondents were taking some or other measure to deal with these changes.

Among the reasons analysed to deal with water paucity, a major one included, 'performing religious rituals to request for rain'. This has at times become the narrative of government institutions in Bhutan too. The participants in these rallies involve a large crowd often more than hundreds of monks and civil servants. They would walk carrying volumes of Buddhist scriptures on their backs, along the main river. The gatherings included representatives and members from a maximum number of communities. According to government officials from Punakha district, they have been performing this ritual for the last six consecutive years.

The other measures dealt with solutions like, developing new, or modifying existing water sharing arrangements between households and villages, improving irrigation channels, shifting from irrigated to rain-fed crops, purchasing irrigation water from residents of upstream and buying electric water pump.

Some other community sharing and selling of water also takes place. When farmers lack their share of the irrigation water, they depend on upstream farmers for irrigation, for which they would provide financial compensation to the upstream farmers.[61]

Conclusion

Bhutanese society has a strong feeling for community. This can also be seen in the ways they perceive as well as preserve nature. In some parts of Bhutan, the rivers are adopted by school students, such cultivation of feeling of belongingness among the young minds represents the strong commitment of the society towards rivers. With increasing industrialization

---

[60]Allison, E. (2004). Spiritually Motivated Natural Resources Protection. In K. Ura & S. Kinga (Eds.), *The Spider and Piglet* (Thimpu: CBS).

[61]Kusters, K., & Wangdi, N. The Costs of Adaptation: Changes in Water Availability and Farmers' Responses in Punakha District. Bhutan. *International Journal of Global Warming*, *V*(IV), 387–399.

74 R. KHARAT AND A. MUNDRA

and urbanization in Bhutan, the reports of water scarcity and depleting quality have been on the rise. This internal water issues can't be sufficed from the groundwater extraction, because of the tough terrain of Bhutan's geography. There are also reports of internal disputes due to unavailability of water resources. This may be due to….

> (T)he quality of metrological data is poor and the available information do not cover the whole country. Though, the data Bhutan has, is insufficient, but then the expertise to gain more data and analyse it, is also lacking in Bhutan. There is a need to raise public awareness, work for sector-specific responses, develop sound coping mechanisms and strengthen local community's assets and knowledge to deal with the adverse impacts of climate change.[62]

This comes as an irony for a water rich country like Bhutan, that it has been marred with a resource crunch in which it is abundant. The issue has been an election question and was also discussed in the National Parliament (*Tshongdu*) of Bhutan.

While, national issue erupted as a question of water availability round the clock, government initiated drinking water policy to check the concerns of common masses. Water has also been a major reason for natural disasters in Bhutan, GLOFs has created many havocs and more such potential glaciers were also identified by scientists. The impact of natural issues is already accentuated by ills of climate change, such that tropical diseases like Malaria and Dengue are found in Bhutan along with scarcity of clean drinking water.

The Hydropower potential has been cherished as the most potent economy driver of Bhutan. This has brought Bhutan as a trade neighbour of India, as India is a buyer of Bhutanese power. But along with it, it also got various internal issues of unemployment, environment concerns, social unrest, etc. The damming of numerous rivers in Bhutan also led to sedimentary runoff in the region and flooding in India. The hydropower is much needed for development and clean energy, Bhutan's run-of-river projects are much needed reality.

The overview of the resources is not fixed by any specific ministry and may sometimes create issues institutional incoherence in their governance. While GNH and Happiness index measures are important as the development indicators in Bhutan, but hydropower is seen as a reason for may be

---

[62] Aanehi, M. *Climate Change in Bhutan: Emerging Issues and Perspectives* (Unpublished M.Phil. dissertation). Jawaharlal Nehru University, New Delhi.

WATER ISSUES IN BHUTAN: INTERNAL DISPUTES AND EXTERNAL TENSIONS   75

not reaching the expected level on these indicators. At the same time one cannot over look the dark side of these hydropower projects which compels to uproot those who basically from rural areas. Chunku Bhutia (2018) has rightly observed in her unpublished thesis,

> The Developmental projects like hydro- power projects should not be justified merely due to its economic benefits like increased food production, enhancement of infrastructures (urban and rural), boosting industrial base etc. Similarly, the people protesting against such development initiatives should not focus merely on the negative impacts of such projects in the form of poverty, displacement etc. So, development is essential for any state/country to survive. But what is needed is a change in the approach of both the policy makers and the people at large regarding the issue of hydro- power projects.[63]

Where Bhutan and India cooperate for transboundary cooperation, some internal disputes may create a gap which manifests into non-coherence in the collaborative efforts. Along with dealing on national and international issues, the local perception also plays a role in dealing with the water issues. In the oriental societies like Bhutan, religious rituals are called upon first to deal with any menace marring the present societies. How much they function, there isn't any measurement scale, but bringing the society together for a cause is surely a major utility provided by them. The internal solace by domestic measures and transboundary enactments from laws formulated by national governments and the cooperation agreements deal the present water resources in Bhutan. Bhutan is quoted as a successful example of water sharing between governments, especially in South Asia, where region is infested with various inhibitions. Water as a resource is precious, rising impact of anthropogenic climate change may raise issues which haven't been of contention yet.

---

[63]Chunku, B. (2018). *Hydro-Power Projects Induced Displacement In Eastern Himalayas: A Comparative Study Of Bhutan and Sikkim* (Unpublished Doctoral thesis). Jawaharlal Nehru University, New Delhi.

The Politics and Policies of Regional Water Management in Southern China

Kris Hartley

INTRODUCTION

Hong Kong has relied for several decades on water supply agreements with neighbouring Guangdong Province in China, and despite halting efforts at self-sufficiency (e.g. alternative source technologies and demand management) the city relies principally on imported supply. This arrangement underscores the importance of maintaining a collaborative relationship within the region, which can be seen as a resilience strategy amidst potential supply threats. This chapter examines water supply threats as theoretically illustrative of the challenges of transboundary governance—the working relationships of government units and agencies within and between cities. The nature of the border between Hong Kong and China is more complex than that of international and intra-national (state or provincial) boundaries. A politically contentious federalism has led to a delicate combination of limited autonomy and implicit subordination. Within this unique setting—described as *one country-two systems*—the strategy of regional collaboration illustrates the connection between governance and resilience. This

K. Hartley (✉)
Department of Asian and Policy Studies, The
Education University of Hong Kong, Tai Po, Hong Kong
e-mail: hartley@u.nus.edu

© The Author(s) 2020
A. Ranjan (ed.), *Water Issues in Himalayan South Asia*,
https://doi.org/10.1007/978-981-32-9614-5_4

77

78 K. HARTLEY

chapter therefore aims to be relevant to environmental management (e.g. climate change adaptation) at both the local and regional scales. To investigate how environmental challenges are addressed through intergovernmental action, this chapter engages the *new institutional economics* literature. This literature addresses the self-interested behaviour of actors within the formal and informal institutions governing interactive spaces. A variety of empirical studies has utilized institutional theory: United States congress members and the role of institutions in collective action,[1] the ability of trust, reputation and reciprocity to facilitate co-management of common pool resources,[2] cooperative equilibrium and coordination within social institutions,[3] and the role of institutions in minimizing transaction costs among contracting parties.[4] A related thread of public choice literature addresses governance failures and inefficiencies that arise from perverse incentives.[5] Theoretical topics addressed by this literature include rent seeking, principal–agent dynamics, the influence of interest groups in democratic systems, and the influence of politics in bureaucratic choice. The literature's concepts of shared resources and coordination are relevant to this chapter's focus on intergovernmental relationships and recurrent collaboration.

In this chapter, the concept of rational choice is scaled up from the individual level to describe the behaviour of government agencies within a transboundary action arena. To achieve this, the chapter utilizes a hybrid analytical framework derived from Ostrom's Institutional Analysis and Development (IAD) framework and Jensen and Lange's transboundary

---

[1] Hall, P. A., & Taylor, R. C. R. (1996). Political Science and the Three New Institutionalisms. *Political Studies, 44*(5), 936–957.

[2] Ostrom, E. (1998). A Behavioral Approach to the Rational Choice Theory of Collective Action: Presidential Address, American Political Science Association, 1997. *American Political Science Review*, 1–22.

[3] Calvert, R. (1995). The Rational Choice Theory of Social Institutions: Cooperation, Coordination, and Communication. *Modern Political Economy: Old Topics, New Directions*, 216–268.

[4] Williamson, O. E. (1981). The Economics of Organization: The Transaction Cost Approach. *American Journal of Sociology*, 548–577.

[5] Self, P., & Peacock, A. (1993). *Government by the Market?: The Politics of Public Choice*. Wiley Online Library; Buchanan, J. M., & Tollison, R. D. (1984). *The Theory of Public Choice—II*. University of Michigan Press; Schwartz, H. M. (1994). Public Choice Theory and Public Choices Bureaucrats and State Reorganization in Australia, Denmark, New Zealand, and Sweden in the 1980s. *Administration & Society, 26*(1), 48–77.

water governance framework. This synthesis combines the treatment of organizations as individual actors with the treatment of transboundary dynamics as institutional parameters regarding a common pool resource. The approach strengthens efforts to study threats in cross-jurisdictional governance contexts by using a language and theoretical basis that harmonizes the two settings. Environmental management often occurs within such settings, justifying the focus on intergovernmental relations to understand resilience.

This chapter proceeds as follows. A case description of Hong Kong and Guangdong Province provides the contextual background for the chapter's later examination of institutional theories and practical policy action. The subsequent section examines scholarly ideas around new institutional economics, in particular the behaviour of public agencies as actors and the dynamics of transboundary relationships. The chapter then applies Hartley's[6] analytical framework to examine transboundary water governance and the role of actor-agencies within a broader regional context, followed by the introduction of a theoretical model describing how Hong Kong's chronic underinvestment in alternative sources has led to a moral hazard. The analysis and discussion section applies aspects of the Hartley framework to understand original interview and official data concerning water management. The conclusion synthesizes the lessons of the chapter and identifies areas for further consideration.

WATER MANAGEMENT IN HONG KONG AND THE PEARL RIVER DELTA REGION

The case of water management dynamics between Hong Kong and China provide an instructive setting in which to explore transboundary governance dynamics. This section describes the recent history of water sourcing in Hong Kong with a focus on collaborative models. China's Dong River is currently the source for 70–80% of Hong Kong's consumable water, and is legally guaranteed through supply contracts with Guangdong province; the province lies immediately adjacent to Hong Kong and encompasses

---

[6]Hartley, K. (2017). Environmental Resilience and Intergovernmental Collaboration in the PRD. *International Journal of Water Resources Development, 34*(4), 525–546.

80 K. HARTLEY

the majority of the river catchment area.[7] The water is sourced from catchments on the mainland side and delivered through an 80-km piping system to Hong Kong for final quality treatment. Hong Kong enjoys a lump-sum package deal that now amounts to 820 million cubic meters (mcm) of water supply per year. A 1989 agreement stipulated a ceiling of 1100 mcm but Hong Kong has not yet approached that level in its sourcing needs. Hong Kong's contractual arrangement is not unusual, as transboundary water agreements are common across mainland China; roughly 40 are currently in place.[8]

Sourcing water from the mainland has an extended history in Hong Kong, beginning with a 1960 agreement for Guangdong province to pipe 22.7 mcm per year from the Shenzhen Reservoir (with the Dong River as the principal source). In 1963, a drought visited substantial hardship on Hong Kong, underscoring the necessity of strengthening water supply reliability. Rationing was introduced to manage demand but negatively impacted public health and agricultural activity. With the failure of stop-gap measures to quickly increase supply, a longer-term effort was made to fashion two freshwater reservoirs from seawater coves and further develop systems for rainwater catchment. Nevertheless, it was clear that a majority of Hong Kong's water would still need to come from the mainland. In the five decades since agreements were introduced, supply has increased tenfold while efforts to develop alternative supply systems stalled or met significant challenges.

As early as the 1970s, researchers warned of an imminent water scarcity crisis in Hong Kong.[9] Despite these and other warnings, in 1982 the Hong

---

[7]China Water Risk. (2014). 8 Things You Should Know About Hong Kong Water. *China Water Risk*. Accessed May 7.

[8]Chen, H., Rieu-Clarke, A., & Wouters, P. (2013). Exploring China's Transboundary Water Treaty Practice Through the Prism of the UN Watercourses Convention. *Water International, 38*(2), 217–230.

[9]Aston, A. (1977). Water Resources and Consumption in Hong Kong. *Urban Ecology, 2*(4), 327–353.

THE POLITICS AND POLICIES OF REGIONAL WATER ...   81

Kong government abandoned the energy-intensive process of desalination,[10] due principally to concern about rising energy costs.[11] The decommissioned Lok On Pai desalination plant, Hong Kong's only such plant in operation at the time, was plagued with inexperienced management and engineering faults, and may have been deemed unnecessary after the completion of a reservoir and unusually high rainfall in 1977 and 1978.[12] Aside from these issues, the fundamental legitimacy challenge to maintaining plant operations was the presence of comparatively inexpensive imported water, rendering desalination a costly alternative. A 2008 paper from the Water Supplies Department (WSD) insisted that a desalination plant was not needed for another 20 years, arguing that supply from the Dong River was adequate.[13] However, Hong Kong is currently exploring a plan to restart desalination, with a reverse-osmosis facility at Tseung Kwan O projected to open in 2022 and expected to produce the equivalent of 49% of the amount of rainwater harvested in 2011.[14] The plant project has recently received support from the government, and studies about financial feasibility, efficiency measures and local ecological impacts were conducted by an external consultancy.

Despite de-emphasizing alternative supply sources and focusing on a single source,[15] Hong Kong has a history of supply innovations, particularly during times of scarcity. The city has operated a seawater toilet flushing program since 1957, with coverage for 85% of the population (272 mcm

[10]Edwards, M. (2013). Hong Kong Water Shortage: A Social & Security Risk. *Austcham News*, No. 157.

[11]Semiat, R. (2008). Energy Issues in Desalination Processes. *Environmental Science & Technology*, 42(22), 8193–8201; Dolnicar, S., & Schäfer, A. I. (2009). Desalinated Versus Recycled Water: Public Perceptions and Profiles of the Accepters. *Journal of Environmental Management*, 90(2), 888–900.

[12]Mody, A. (1997). *Infrastructure Strategies in East Asia: The Untold Story*. World Bank Publications.

[13]Eng, D. (2008, April). Desalination 'Not Needed for at Least 20 Years'. *South China Morning Post*. http://www.scmp.com/article/633711/desalination-not-needed-least-20-years.

[14]HKLC. (2012). *345WF—Planning and Investigation Study of Desalination Plant at Tseung Kwan O, Supplementary Information*. Hong Kong: Hong Kong Legislative Council Panel on Development. http://www.legco.gov.hk/yr11-12/english/panels/dev/papers/dev0417cb1-1855-1-e.pdf.

[15]Liu, S. (2014, January 13). Water: Tale of Two Cities. *China Water Risk*. http://chinawaterrisk.org/opinions/water-tale-of-two-cities/.

82   K. HARTLEY

per year) through separate infrastructure and plumbing systems in most buildings. Hong Kong has recently embarked on a plan to increase the coverage of seawater flushing systems to satellite towns and remote areas. Sea-bound freshwater reservoirs were also a significant innovation, with Hong Kong's two being the world's largest at the time of their opening.

Over the past several decades, water demand in Hong Kong has tracked the city's population and economic growth; aggregate demand is now roughly 1200 mcm per year. Restructuring from manufacturing to service industries has reduced industrial demand, partially offsetting increases in demand from population growth (0.7% yearly) and the intensive usage patterns typical of wealthier households.[16] The Hong Kong government projected the city's population to increase by as many as 700,000 residents between 2010 and 2020, or 70,000 per year,[17] while the Hong Kong WSD projected a 1.5 million increase between 2012 and 2040, or 53,500 per year.[18] Aggressive estimates project that by 2030, Hong Kong's water demand may be 40% higher than the 2012 level.[19] In this scenario, Hong Kong's reliance on Dong River water may be as high as 1320 mcm per year, well in excess of the current 1100 mcm ceiling. In a lower-growth scenario, Hong Kong's demand for Dong River water may eventually be only 713 mcm per year, below current withdrawal levels (with alternative supply more than offsetting demand growth). Both projections are based on the assumption that the Dong River will eventually contribute 41–56% of Hong Kong's future water supply (an average of 48%), and the lower

[16]Mui, K. W., Wong, L. T., & Law, L. Y. (2007). Domestic Water Consumption Benchmark Development for Hong Kong. *Building Services Engineering Research and Technology, 28*(4), 329–335.

[17]HKLC. (2012). *345WF—Planning and Investigation Study of Desalination Plant at Tseung Kwan O, Supplementary Information.* Hong Kong: Hong Kong Legislative Council Panel on Development. http://www.legco.gov.hk/yr11-12/english/panels/dev/papers/dev0417cb1-1855-1-e.pdf.

[18]Water Supplies Department (WSD). (2008). *Total Water Management in Hong Kong: Towards Sustainable Use of Water Resources.* Hong Kong Water Supplies Department (WSD), Government of the Hong Kong Special Administrative Region.

[19]Chan, M. (2013, March 22). Hong Kong's Unsustainable Water Policies. *South China Morning Post.* http://www.scmp.com/comment/insight-opinion/article/1196453/hong-kongs-unsustainable-water-policie.

THE POLITICS AND POLICIES OF REGIONAL WATER ... 83

projection also assumes that freshwater demand (net of seawater flushing) will rise to 1100 mcm.[20]

From a regional perspective, Hong Kong's water sustainability is reliant also on demand patterns within neighbouring Guangdong province. More than 40 million residents in the greater Pearl River Delta (PRD) region rely on water from the Dong (East), Bei (North), Xi (West), and Pearl Rivers. The long-term aggregate demand profile for Dong River water is uncertain due to growing household and industrial water demand in China's upstream catchments.[21] For example, Guangdong province embarked on a plan to increase power generation capacity by 45% between 2010 and 2015 in order to accommodate 15.9 million new inhabitants; population growth at this pace strains water supply.[22] Further, several of China's most economically vibrant cities (Guangzhou, Shenzhen, Dongguan and Huizhou) rely on the Dong River for water supply, while upstream cities (Heyuan, Meizhou and Shaoguan) likewise make claims on Dong River water. Some of these cities are projected to face significant water shortages by 2020.[23] Nevertheless, the Guangdong region is evolving into a more innovation and knowledge-driven economy, with a focus on higher-value goods.[24] Therefore, demand for water may not grow as rapidly as in recent decades due to declining industrial demand. Population growth is also expected to flatten in the coming decades and has already seen a recent trend towards slowing growth.[25]

[20]Water Supplies Department (WSD). (2008). *Total Water Management in Hong Kong: Towards Sustainable Use of Water Resources*. Hong Kong Water Supplies Department (WSD), Government of the Hong Kong Special Administrative Region.

[21]Chan, M. (2013, March 22). Hong Kong's Unsustainable Water Policies. *South China Morning Post*. http://www.scmp.com/comment/insight-opinion/article/1196453/hong-kongs-unsustainable-water-policie; Liu, S., & Williams, J. (2014). *The Water Tales of Hong Kong and Singapore: Divergent Approaches to Water Dependency*.

[22]LeClue, S. (2012). Hong Kong: Stepping Up Water Security? *China Water Risk*. http://chinawaterrisk.org/resources/analysis-reviews/hong-kong-stepping-up-water-security/.

[23]Liu, S. (2012, July 11). A Vulnerable Dongjiang Is a Vulnerable HK. *China Water Risk*. http://chinawaterrisk.org/opinions/a-vulnerable-dongjiang-is-a-vulnerable-hong-kong/.

[24]Asia Business Council. (2011). *Economic Transformation of the Greater PRD: Moving Up the Value Chain*. Asia Business Council. http://www.asiabusinesscouncil.org/docs/PRDBriefing.pdf.

[25]Since 2010, annual population growth in Guangdong province has been less than 1%. http://www.stats.gov.cn/english/.

84 K. HARTLEY

INSTITUTIONAL ANALYSIS AND AGENCIES AS ACTORS

To proceed with an institutional analysis of the water management case, it is necessary to consider the academic literature regarding institutional analysis. The institutional environment in the PRD region can be viewed through two theoretical lenses: public agencies as actors and complex institutional conditions regarding territorial jurisdictions. The rational choice model used in this chapter extends the analytical scope from the individual to the organizational level, under the assumption that organizations as individual entities can exhibit the same types of humanly characteristics—foresight, ambitions and biases—that motivate behaviour in individual persons. This approach also implies that theories about rational behaviour, which inform studies about individuals, can be extended to explain the behaviour of organizations under given institutional conditions. Emerging from work done by neoclassical game theorists[26] and cognitive psychologists,[27] models of behaviour that assume rationality under the assumptions of complete information and self-interest can help explain patterns of collaboration and reciprocity among agencies (e.g. for management of natural resources). Scholarship critical of the assumptions underlying rational theories of behaviour[28] is likewise useful. Such assumptions, including rationality itself, have been discredited in some theoretical circles, leading not only to adaptations of the original model (as in Simon's notion of *bounded rationality*) but also to a complete repudiation of behavioural modeling as a scholarly practice.

While there is continuing debate about the certitude of behavioural theory assumptions, few studies explicitly analogize the behaviour of individual users within a common pool resource dilemma to that of organizations in

---

[26] Harsanyi, J. C. (1977). Morality and the Theory of Rational Behavior. *Social Research*, 623–656; Gintis, H. (2000). Beyond Homo Economicus: Evidence from Experimental Economics. *Ecological Economics*, 35(3), 311–322; Axelrod, R. (1984). *The Evolution of Cooperation*. New York: Basic Books; Robbins, L. (1935). *An Essay on the Scope and Nature of Economic Science*. London: Macmillan; Binmore, K. G. (1998). *Game Theory and the Social Contract: Just Playing* (Vol. 2). MIT press.

[27] Kahneman, D., & Tversky, A. (1979). Prospect Theory: An Analysis of Decision Under Risk. *Econometrica: Journal of the Econometric Society*, 263–291; Simon, H. A. (1957). *Models of Man: Social and Rational*. Oxford: Wiley.

[28] Ostrom, E. (1998). A Behavioral Approach to the Rational Choice Theory of Collective Action: Presidential Address, American Political Science Association, 1997. *American Political Science Review*, 1–22.

similar circumstances. Bushouse[29] explores the behavior of competing firms in for-profit and non-profit sectors through the IAD's constitutional level of analysis, but the study's focus on club goods does not reflect this chapter's conceptualization of water as both a toll good and common pool resource. The economy of public goods has also been explored through the concept of polycentricity,[30] including multi-organizational arrangements for managing ecologically protected areas.[31] In such studies, the behaviour of organizations and agencies is the unit of analysis, offering opportunities to more deeply understand and theorize institutional conditions. For example, Araral's[32] examination of the strategic interaction between aid donors and recipients examines public agencies and donor organizations as actors. In his model, the goal of the bureaucracy is survival, while the goal of the donor agency is to expand its own loan portfolio. Araral frames the behaviour of these actors as products of incentive problems and moral hazard, a theme this chapter explores in the context of water.

Ostrom's research on Socio-Ecological Systems (SES) builds on the IAD to describe the dynamics of self-organization in addressing common pool resource challenges.[33] The SES framework focuses on the interplay of four subsystems, the most directly relevant of which is governance. Two others applicable to transboundary water management are resource systems and resource units, while the fourth—users—is subsumed under governance for analysis of actor agencies. Individual variables of the SES framework are also applicable to this case of transboundary environmental management, including attributes of the resource system, resource unit mobility,

---

[29] Bushouse, B. K. (2011). Governance Structures: Using IAD to Understand Variation in Service Delivery for Club Goods with Information Asymmetry. *Policy Studies Journal, 39*(1), 105–119.

[30] McGinnis, M. D. (1999). *Polycentricity and Local Public Economies: Readings from the Workshop in Political Theory and Policy Analysis.* University of Michigan Press.

[31] Oakerson, R. J., & Parks, R. B. (2011). The Study of Local Public Economies: Multi-organizational, Multi-level Institutional Analysis and Development. *Policy Studies Journal, 39*(1), 147–167.

[32] Araral, E. (2009). The Strategic Games That Donors and Bureaucrats Play: An Institutional Rational Choice Analysis. *Journal of Public Administration Research and Theory, 19*(4), 853–871.

[33] Anderies, J. M., Janssen, M. A., & Ostrom, E. (2004). A Framework to Analyze the Robustness of Social-Ecological Systems from an Institutional Perspective. *Ecology and Society, 9*(1), 18.

86   K. HARTLEY

presence of rules across types (operational, collective choice and constitutional), and characteristics of user-actors. Resource system attributes can be used to analyse diverse geographic contexts, while unit mobility is relevant to water settings where interlinked delivery infrastructures connect source regions with high-demand regions downstream. While such frameworks have been valuable contributions to institutional analysis for environmental management, the concurrent classification of water as both a toll good and common pool resource captures the complexity of the Hong Kong governance setting in a way unaddressed by existing frameworks.

Finally, it is necessary to consider the literature about governance dynamics across borders. Such topics in the Hong Kong-Guangdong region have received ample scholarly attention, particularly after the 1997 handover and in the wake of rapid industrialization throughout the PRD region. Given the *one country-two systems* approach—an ambiguous arrangement that eludes comparison—scholarship has characterized the Hong Kong-China relationship in various ways: 'constitutionalism with partial democracy',[34] the paradox of 'symbiotic interactions' and 'separation of systems',[35] and federalism.[36] Hills[37] examines the relationship between Hong Kong and China through the perspective of regional environmental management, arguing that Hong Kong has been slow to develop coordinated policies but that objectives shared with the mainland may ultimately enable closer cooperation on 'ecological modernization'. Regarding transboundary initiatives and the influence of scale, Lee[38] compares the regime approach of

[34] Chen, A. H. Y. (2009). The Theory, Constitution and Practice of Autonomy: The Case of Hong Kong. In *One Country, Two Systems, Three Legal Orders-Perspectives of Evolution* (pp. 751–767). Springer.

[35] Holliday, I., Ngok, M., & Yep, R. (2004). After 1997: The Dialectics of Hong Kong Dependence. *Journal of Contemporary Asia, 34*(2), 254–270.

[36] He, B., Galligan, B., & Inoguchi, T. (2009). *Federalism in Asia.* Edward Elgar; Waldron, A. (1990). Warlordism Versus Federalism: The Revival of a Debate? *The China Quarterly, 121,* 116–128; Harding, H. (1993). The Concept of 'Greater China': Themes, Variations and Reservations. *The China Quarterly, 136,* 660–686.

[37] Hills, P., & Roberts, P. (2001). Political Integration, Transboundary Pollution and Sustainability: Challenges for Environmental Policy in the PRD Region. *Journal of Environmental Planning and Management, 44*(4), 455–473.

[38] Lee, N. K. (2013). The Changing Nature of Border, Scale and the Production of Hong Kong's Water Supply System Since 1959. *International Journal of Urban and Regional Research.*

political-based localization to the political ecology perspective of resource-based localization, adding 'critical' border studies to analyses of water management.

In a comparative case, Ho and So[39] examine borderland dynamics and integration in Hong Kong and Singapore. According to the authors, the Hong Kong-Guangzhou borderland region is more 'culturally contiguous' with respect to language, customs and bonds at the family and community level. Another distinguishing factor is the flow of investment capital, which is two-directional between Hong Kong and Guangdong. According to the authors, the Chinese side of the borderland has experienced industrialization and will continue its rapid development, while Hong Kong maintains a shift to lower water-intensive economic growth (e.g. services). This adds complexity to the transboundary dynamic, as Guangdong's cities are helping sustain China's economic growth in ways that are more water-intensive than those in Hong Kong. At the same time, Hong Kong is seen as an important economic engine for China, and this symbiotic relationship shapes the regional water governance setting in both implicit and explicit ways. While this chapter focuses on systemic and administrative differences between the two sides of the border, it also acknowledges political sensitivities. Indeed, the studies cited in this review, and many others, represent useful efforts to capture political dynamics in transboundary relationships. This chapter proceeds by describing an analytical framework that captures the institutional conditions shaping the transboundary governance of natural resources and other borderless phenomena.

Analytical Framework

At a fundamental level, institutional frameworks exist to systematize studies of complex systems in which multiple actors with various interests, resource endowments and degrees of power interact under a mutually agreed set of norms, rules and precepts. The scales in which institution-focused analytical frameworks have been applied to natural resource governance vary widely, from river basins and irrigation plains to entire regions with fragmented and multi-layered governance systems. Frameworks are designed to capture all elements of complexity within a collaborative situation, but the explicit treatment of transboundary dynamics is lacking. This oversight limits the

---

[39] Ho, K. C., & So, A. (1997). Semi-periphery and Borderland Integration: Singapore and Hong Kong Experiences. *Political Geography, 16*(3), 241–259.

explanatory power of frameworks in examining settings such as the PRD, Nile River, Johor watershed in Singapore and Malaysia, and the Colorado River in the United States and Mexico. Further refinement of analytical frameworks is needed also to capture the complex characteristics of natural resources in shared settings. In particular, the distinction between toll goods (low cost of exclusion and non-rivalrous consumption) and common pool resources (high cost of exclusion and non-rivalrous consumption) has implications for the implementation of formal contracts and the role of trust and reciprocity in collaborative settings. In many cases, these two factors—transboundary dynamics and diversity in good type—combine to generate empirical complexity that eludes standard frameworks.

Actor Agencies

This theoretical proposal seeks to explain how agency-actors pursue organizational objectives within the institutional structure of transboundary settings. Individual actors are agencies whose characteristics within the framework resemble those of individual persons. Ostrom's four clusters of variables are a useful guide for analysing these characteristics in a comparison between individuals and agencies (Table 1). At a fundamental level, self-interest underlies the perceptions and behaviours of actors, so this analysis focuses on self-interest in three dimensions: origins, strategic execution and impact. These dimensions correspond roughly with the policy process,

Table 1 Comparing individuals and agencies in institutional analysis

Variable cluster	Individual	Agency
Resources	Property; personal influence	Budget; assets; legal power
Basis of valuation for conditions and actions	Ambitions, preferences and biases	Strategic organizational and national objectives
Analytical capacity	Education; access to information processing capabilities	Internal analytical units; capabilities of employees
Action decisions	Cognitive processes; degree of rationality; impacts on others	Internal policies; precedent; legal options; impacts on stakeholders and intergovernmental relationships

Source Author, adapted from Ostrom's IAD

namely the chronological flow of initiatives from conception to implementation.[40] They also relate respectively to the three analytical dimensions of the IAD: conditions, action arenas and outcomes. First, the origins of agency interests are determined largely by hierarchies and power structures; agencies must support the strategic directives of parent entities (agencies and governments). For this reason, the individual strategies of agencies can appear to be insular (narrowly focused), self-serving (exclusion of collective welfare) and mutually contradictory in regional settings, as political or jurisdictional boundaries represent the limits of agency accountability.

Second, coordination of top-down strategic initiatives can be weak where resource management is divided into functional areas across multiple agencies. Agencies may have varying levels of autonomy in executing strategic directives, depending on their status in the administrative hierarchy. As such, governance bottlenecks and related inefficiencies arise when management of the resource life cycle, as from origin to consumption, occurs across agencies and with differing levels of flexibility and autonomy. As a policy moves from its strategic genesis to execution point, a phenomenon resembling distance decay occurs in the effectiveness of coordination; such dynamics are commonly addressed in literature about policy implementation and street-level bureaucrats.[41] Depending on the individual behaviours of agencies, heterogeneous adherence to distributed policy directives contributes to lapses in coordination and weakens central strategic control.

Third, the degree of agency discretion becomes relevant as differing interpretations of the same strategic initiative lead to varying outputs. This relates to the impact dimension of the IAD. Even clear strategies, when interpreted by individual agencies, can produce differing and potentially conflicting approaches to execution, and thereby suboptimal, inefficient, or mutually contradictory outcomes. As such, the rational choice model holds

---

[40]See Lasswell, H. D. (1951). The Policy Orientation. In *The Policy Sciences* (pp. 13–14). Stanford: Stanford University Press; Jann, W., & Wegrich, K. (2007). Theories of the Policy Cycle. In *Handbook of Public Policy Analysis* (p 43); Howlett, M., Ramesh, M., & Perl, A. (1995). *Studying Public Policy: Policy Cycles and Policy Subsystems* (Vol. 163). Cambridge University Press.

[41]Lipsky, M. (2010). *Street-Level Bureaucracy, 30th Ann. Ed.: Dilemmas of the Individual in Public Service*. Russell Sage Foundation; Sabatier, P. A. (1986). Top-Down and Bottom-Up Approaches to Implementation Research: A Critical Analysis and Suggested Synthesis. *Journal of Public Policy*, 6(1), 21–48; Meyers, M. K., Vorsanger, S., Peters, B. G., & Pierre, J. (2003). *Street-Level Bureaucrats and the Implementation of Public Policy*. London: Sage.

that individual agencies act not only within the authoritative confines of higher strategic directives but also within their own interest in operational efficiency and strategic efficacy. Outcomes may be shaped less by the quality of strategic initiatives and implementation capabilities of agencies than by overarching governance structures and related coordination mechanisms.

Understanding Institutional Dynamics

This discussion now turns to transboundary dynamics as institutional parameters. In this chapter's framework, the sphere of interaction is defined by geography, relationships and strategic options. Conditions governing such situations are already captured by the IAD (Fig. 1). In the environmental context, physical condition refers to natural laws governing the good of concern, including its extraction, use, and relationship to the broader physical setting. In describing the physical and material characteristics of a good, Ostrom utilizes a 2 × 2 matrix overlaying cost of excludability (denying non-paying parties access) with degree of rivalry (zero-sum consumption).[42] A good with high cost of excludability and high degree of rivalry is a common pool resource; one with low cost of excludability and

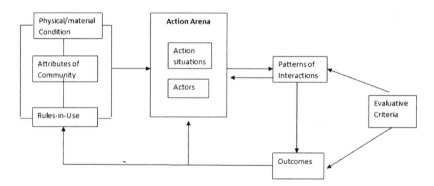

Fig. 1 Institutional analysis and development framework (Ostrom, E. [2007]. Institutional Rational Choice: An Assessment of the Institutional Analysis and Development Framework)

---

[42] Polski, M. M., & Ostrom, E. (1999). *An Institutional Framework for Policy Analysis and Design* (Workshop in Political Theory and Policy Analysis Working Paper W98-27). Bloomington: Indiana University.

THE POLITICS AND POLICIES OF REGIONAL WATER ... 91

high degree of rivalry is a toll good. This examination of water as toll good accounts for formal water procurement contracts between transboundary parties.

Attributes of the community include generally accepted norms of behaviour, shared understanding about issues, preferential homogeneity, and the role of trust, reciprocity and reputation. Rules-in-use relate to boundary demarcation, bargaining positions, scope of authority and the production and circulation of information, among other factors relevant to the design and execution of agreements and contracts. Given these conditions, actors come together within the action arena to develop solutions to resource management problems. In this analytical context, the characteristics of actors (resources, preferences, etc.) and those of the situation (participants, outcome possibilities, costs and benefits, etc.) should be systematically appraised. A crucial assumption in this exercise is that the institutional setting is uniform in its conditions and characteristics, allowing analytical efforts to focus on patterns of interactions and outcomes while holding contextual variables constant where possible. However, variations in institutional settings across the study area often arise in more complex and heterogeneous settings, such as those involving transboundary collaborations and agreements. This requires an extension of the IAD framework.

Jensen and Lange[43] propose a framework with dimensions that, in combination with the IAD, have the potential to capture the complex institutional dynamics of transboundary settings (Fig. 2). The authors' framework identifies stakeholder types in the overall political economy of water, including private investors, donors, civil society organizations (NGOs) and intergovernmental organizations such as regional cooperatives. The authors also cite 'riparian governments' such as water and environmental agencies and those overseeing activities with environmental impact (e.g. mining, energy and agriculture). The Jensen and Lange framework facilitates the analysis of political and institutional dimensions in multi-layered governance settings. Most relevant to this synthesis, the framework accounts for the influence on broader development strategies by stakeholders within national and

---

[43]Jensen, K., & Lange, R. (2013). *Transboundary Water Governance in a Shifting Development Context* (DIIS Report 2013:20). Danish Institute for International Studies. https://www.diis.dk/en/research/transboundary-water-governance-in-a-shifting-development-context-0.

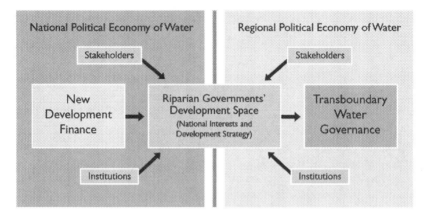

Fig. 2 Framework for transboundary water governance (Jensen, K., & Lange, R. [2013]. *Transboundary Water Governance in a Shifting Development Context* [p. 25, DIIS Report 2013:20]. Danish Institute for International Studies. https://www.diis.dk/en/research/transboundary-water-governance-in-a-shifting-development-context-0)

regional spheres. As such, it introduces a useful meta-scalar dimension. In examining transboundary challenges, it is useful to extend the scope of analysis beyond political boundaries and the jurisdictions of individual agencies.

Figure 3 is a framework that modifies the IAD to capture transboundary dynamics.[44] The framework accounts for the local and regional (transboundary) scale, and for all analytical elements (physical condition, rules-in-use and attributes) proposed by the original form of the IAD. The value of this approach is that the content of the three analytical elements is unique to the scale under consideration, generating opportunities for richer analysis. The theoretical value is more explicit identification of variables and relationships explaining the behaviour of actors with diverse interests and varying levels of influence, both within and across political boundaries. The flexibility of the framework allows it to concurrently capture transboundary dynamics such as agreements and collaborations common to natural resource management.

---

[44] Hartley, K. (2017). Environmental Resilience and Intergovernmental Collaboration in the PRD. *International Journal of Water Resources Development*, *34*(4), 525–546.

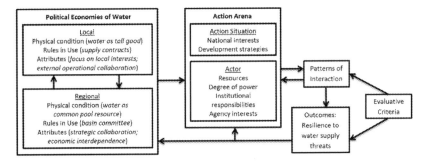

Fig. 3 Hybrid framework (Hartley, K. [2017]. Environmental Resilience and Intergovernmental Collaboration in the PRD. *International Journal of Water Resources Development*, 34[4], 525–546)

The framework assumes the meta-structure of the IAD framework, but with the section originally devoted to physical condition, rules-in-use, and attributes relabelled 'political economics of water' and split into local and regional scales. The Jensen and Lange framework is also absorbed into the IAD's 'action situation', dividing the analysis into interests and strategies that account for individual and collective preferences. In addition to reducing institutional complexity into a two-scale framework, this approach provides an analytical tool for comparing domestic governance traits across international settings.

Understanding Investment in Alternative Sources

In addition to understanding institutional dynamics, it is helpful to frame the theoretical basis for Hong Kong's water management strategy. Figure 4 illustrates how the long-running dependence of Hong Kong on Dong River water leads to chronic underinvestment in alternative supply capacity. Such a situation is described by theory as a moral hazard, which in this case exposes Hong Kong to fluctuations in price, continuity and quality of water supply and de-incentivizes efforts to build redundant or substitute capacity.

B_1 represents the current budget for water procurement (Dong River contracts and alternative sources). B_2 represents a hypothetically increased budget for procurement, funded by higher tariffs and gains in usage efficiency. Q_i represents the fixed amount of water supply from the Dong River as stipulated in transboundary contracts; this amount does not vary based

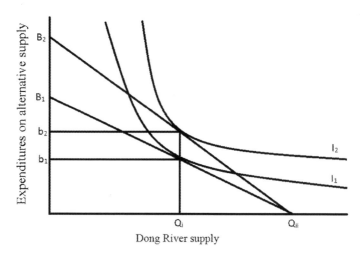

Fig. 4 Investment in alternative supply (*Source* Author)

on changes in Hong Kong's overall budget for procurement. Q_{ii} represents a hypothetical fixed point for the quantity of water supplied to consumers if no expenditures were appropriated to supply from alternative sources. I_1 and I_2 represent the respective indifference curves between expenditures for Dong River supply and for alternative sources. Gains realized from a change from B_1 to B_2 would be applied largely to expenditures on alternative supply (b_1–b_2), assuming continuation of the Dong River contract terms in volume and price. As such, proceeds from a tariff levy could be used for the development, operation and maintenance of alternative sources such as desalination and wastewater purification. Over an extended period of time, additional capacity in alternative sources may fill the supply gap caused by increased demand and upstream rivalry for Dong River water, thereby reducing Hong Kong's reliance on imported water. This would have implications not only for the city's supply sustainability and resilience in the event of a Dong River supply shock, but also for any uncertainty arising from constantly evolving political dynamics.

There are signs that this moral hazard may be dissipating. In 2008, Hong Kong's government introduced a Total Water Management (TWM) initiative to address the sustainability challenge from two angles: demand management and water supply stability. TWM[45] is said to serve two purposes: 'to better prepare Hong Kong for uncertainties such as acute climate changes and low rainfall', and 'to enhance Hong Kong's role as a good partner to other municipalities in the PRD in promoting sustainable use of water'. TWM is informed by research about demand and supply patterns, cost effectiveness, environmental impacts and public awareness. It calls for long-term (to 2030) supply sustainability strategies that include protection and conservation of water resources and development of new sources like greywater recycling and desalination. Other initiatives targeted at demand management include the Water Efficiency Labelling Scheme for appliances and faucets, leakage control, promotion of water-saving devices and extension of seawater flushing. Many conservation schemes are expected to be managed under the Water Intelligence Network (WIN), an information system that collects and processes real-time data to support operational and policy decisions. An expanded supply portfolio includes collection of rainwater and greywater, and the processing of sewage into treated effluent used for non-potable uses like street cleaning and landscape irrigation. Additionally, the Tseung Kwan O desalination facility will have a capacity of 50 mcm per year, expandable to 100 mcm. By 2030, TWM expects to save 236 mcm per year, of which 80% would be accounted for by water conservation and active leakage control, and the remainder by seawater flushing and water reclamation. A revision of TWM, underway as of 2018, is expected to include evaluations of current measures, proposed initiatives and forecasts of demand and supply to 2040.

ANALYSIS AND DISCUSSION

The analysis in this section is based on interviews with officials, academics and other representatives from Hong Kong and Guangdong province holding influence over or expertise in water management in the region. For parsimony, findings are interpreted through the three sub-dimensions of the political economy dimension of the framework previously described (Fig. 3: physical condition, rules-in-use, and attributes); the focus is on the regional scale. When viewed through the physical condition sub-dimension, the one-way carriage of water from mainland China to Hong Kong, across an

---

[45]Water Supplies Department (WSD). (2008). *Total Water Management in Hong Kong: Towards Sustainable Use of Water Resources.* Hong Kong Water Supplies Department (WSD), Government of the Hong Kong Special Administrative Region.

administrative border, implies that the former has more leverage than the latter. That is, decisions that mainland China makes about water (related to quality, continuity of supply, and other factors) affect Hong Kong in ways that lack recourse outside of contract stipulations. When viewed through the rules-in-use sub-dimension, the presence of contracts implies that water is a toll good; that is, Hong Kong pays for and receives a good based on negotiated terms. It is in the attributes sub-dimension where the theoretical nuance between local and regional scales emerges and justifies the amendment to the IAD framework, as this sub-dimension captures both institutional factors and intangible norms. Given that the Dong River is the primary supply source for much of the region, its management is crucial not only for Hong Kong but also for mainland China cities. The presence of basin management committees implies that governments in the region understand this necessity. As such, water can be seen not only as a toll-good (in the context of Hong Kong's role) but also as a common pool resource, in which all parties are impacted by the behaviour of all others, including upstream parties with a higher degree of control over the physical condition of water.

A shared understanding about issues affecting all parties, principally water scarcity, quality and distribution, provides a basis for collaboration built on mutual interest. Local governments are generally prevented by law and norms of behaviour from indulging in selfish actions that threaten the welfare of regional peers. Whether by central government edict or by their own volition, governments in the region appear to acknowledge the sensitivity of Dong River supply and are making cooperative efforts to manage it. However, there can be differences in how governments view collaboration and these fall along historical-political lines.

Climate Change

One assumption of the IAD is that actors make strategic adjustments to changing circumstances. The Hong Kong case illustrates two contexts where variable circumstances and uncertainty require strategic adjustment: climate change and political tension. In a climate change scenario, the number of rainy days might be expected to decrease while the intensity of rain increases.[46] This would impact not only Dong River flow but also

---

[46] According to the WSD, Hong Kong already experiences "erratic" rainfall, which has recently fluctuated from 101 mcm in 2011 to 336 mcm in 2013.

local catchment yield. Governments have approached such threats from a technical perspective; more efficient piping and storage systems are complemented by diversification away from sources with high vulnerability. More recently, the wide body of science about climate change is having a deepening influence on policymaking, particularly in the use of evidence-based policy for scenario planning and contingency analysis. This adaptive approach has been empirically observed and is theoretically supported; one example is Ostrom's[47] study of rural areas in developing countries, where flexible institutional arrangements have been used to manage natural disasters. Flexibility to respond to water supply threats as embodied by the strengthening of intergovernmental collaboration is a potential form of resilience for the PRD region.

The legal and operational mechanics of water supply contracts exist between Hong Kong and Guangdong province; due to region-wide reliance on Dong River water, municipalities within the basin also have an interest in these mechanics. Management of the Dong River occurs under the purview of a basin committee comprising representatives from cities and localities on the mainland side. As such, Hong Kong's sustainability relies to some degree on the commitment of mainland localities to maintain supply continuity through demand management and quality through regulation of polluting activities. Beyond contractual arrangements that include water supply and eco-compensation to upstream areas, Hong Kong's softer relationships with upstream municipalities are supported by delegation programs and by the outreach of Hong Kong NGOs to communities in source regions. Environmental conditions, as impacted by urban growth and industrial and agricultural activity in source regions, can affect downstream water quality. Awareness about and efforts to improve these conditions have been promoted by NGOs, with a focus on resilience against extreme weather events related to climate change. Such events can either reduce water flow, thereby raising the concentration of pollutants, or increase water flow, causing erosion and harm to riparian environments. Tighter regulation of human activities, including urban development in watershed areas, is necessary to maintain water quality in a scarcity scenario and to enhance runoff absorption in minimizing the impacts of floods.

---

[47] Ostrom, E. (2011). Background on the Institutional Analysis and Development Framework. *Policy Studies Journal, 39*(1), 7–27.

Political Dynamics

> Is water as resource political? Yes. It is a highly political issue, and without addressing the political situation we cannot address water policy. The most urgent and obvious [concern] is political. It is the most uncertain...[*we*] don't know what [*China*] is going to do.
>
> Independent water expert

Given the salience and complexity of the political situation between Hong Kong and China, a politics-blind assessment of transboundary water management dynamics in the PRD would be incomplete. Water is a resource that is fundamental to economic and social stability—and by extension political legitimacy; thus, an analysis that ignores the politics of water would overlook crucial determinants of resilience. The evolving nature of this case's political dynamics introduces an element of tension to the otherwise apolitical relationship between Hong Kong and mainland China regarding water contracts and supply operations. In Hong Kong, the so-called 'umbrella revolution', 'occupy central' movement (2014), 'fishball revolution' (2016), and 2019 pro-democracy protests were combustible expressions of a movement that seeks in part to preserve the city's political autonomy. These events have increased political tensions between Hong Kong and China, and have also raised questions about how China's central government plans to build and maintain legitimacy in a city that was a British colony until 1997. Hong Kong's reliance on Dong River supply had previously raised concerns that China would use the city's water resource dependence to extract political concessions.[48] While some scholars envision the absorption of 'greater China' into the mainland occurring over time, the memory of autonomy and notion of Hong Kong independence is likely to endure among 'localist' advocates for the foreseeable future. The political situation can therefore be considered uncertain, as may possible strategies by China's central government.

> China has strong governance and can handle things. If there is a problem, get it solved. Government is a machine. In terms of Hong Kong, the arrangement

---

[48]Chau, K. W. (1993). Management of Limited Water Resources in Hong Kong. *International Journal of Water Resources Development, 9*(1), 65–73.

THE POLITICS AND POLICIES OF REGIONAL WATER ... 99

is ok but [*there is*] a political issue. Young people [*are*] promoting localization or independence.

Water expert, Chinese University of Hong Kong

The current political environment makes it more difficult to initiate further collaboration [*between Hong Kong*] and the mainland.

Water expert, University of Hong Kong

The perception of water as a political resource has a history in the relationship between Hong Kong and China. During the 1960s water crisis, China's provision of water to Hong Kong was seen as an essential lifeline to the city and a tacit acknowledgement that China still viewed Hong Kong as part of a larger cultural polity. Cultivating this relationship was also important for maintaining reciprocity, as China later relied on Hong Kong as an entry portal for global investment as the mainland progressively opened to trade.

For Dong River water there [*were*] two terminologies: economic water and political water...in early 1960s, when for three years southern and central China faced an extended drought. Zhou Enlai and Mao Zedong initiated a program. At that time it was political water, because it wasn't a business deal. This was in line with Beijing's [*eventual*] policy in 1997: to keep Hong Kong as it is, and it would benefit mainland development. China was closed to the world, and Hong Kong was the only channel in the 1950s. In the 1960s, China helped Hong Kong by providing water; with the so-called three million brothers and sisters. In the late 1970s, China opened up funding connections and Hong Kong helped China again.

Water expert, Chinese University of Hong Kong

The challenge in Hong Kong is that it doesn't have a high moral ground to demand more water. If a huge drought strikes [*the region*] we would need to reduce consumption. Dong River [*authorities*] can say Hong Kong gets priority, but Hong Kong is in no position to take this. It would have to be responsible [*and*] share. At [*such*] time, it's an ethical challenge if your neighbor is thirsty just because you demand so much water.

Independent water expert

In the context of water management, the case's transboundary setting is more usefully understood as one between Hong Kong and individual governments in the region than between Hong Kong and China's central government. Nevertheless, given political tensions, the focus of activists, media and casual observers tends towards the latter, as it feeds conveniently into narratives about political and systemic dualism of the sort that were used to understand geopolitical power struggles throughout the twentieth century. Although water has been seen as a political resource, the daily realities of water as an economic and social resource are leading local governments to take a pragmatic approach towards supply sustainability. This includes moving beyond water contracts to establish working relationships, trust and reciprocity with the regional and local governments whose strategies materially impact Hong Kong's water security. Further, if China resumes governance decentralization reforms, local governments will become more autonomous, underscoring the value of close intergovernmental ties. With more autonomy, governments might feel free to pursue mutually beneficial arrangements with Hong Kong, independent of simmering political sensitivities. As such, the transboundary relationship from a regional collaboration perspective becomes at once less complex (absent overt political tension) and more complex (number of actors and interdependencies) than the relationship from a nation-to-nation perspective. Indeed, interviewees in Hong Kong's WSD viewed the political situation as a side-issue; their practical goal is to keep water taps flowing now and into the future. Establishing working relationships with cities in the region supports this objective.

Conclusion

Water governance has often been explored from purely hydrological and technical points of view. Infrastructure (pipe diversions and storage) and technology (quality monitoring and enhancement) have complemented demand management efforts aimed at building supply sustainability. However, this case underscores an emerging tool for building resilience: the establishment of regional governance relationships that help individual cities prepare for supply threats. For Hong Kong, this changes water from a toll good to a common pool resource. Regarding climate change, a network of operational interdependencies ensures that responses (e.g. rationing) to extreme weather events are made from a regional perspective, with shared

THE POLITICS AND POLICIES OF REGIONAL WATER ... 101

sacrifice based on trust and reciprocity. This approach reflects the IAD's redistributional equity criteria with respect to shared access, costs and externalities in supplying water. In building resilience against supply shocks and other unforeseen events, close operational collaborations across the border are crucial for diffusing tensions that might surface during a scarcity scenario. Any assumption that Hong Kong can rely on contracts even if upstream supply conditions deteriorate is not based on the reality of the circumstances examined in this study, including uncertainty about water security with respect to climate change.

Combining the IAD and Jensen and Lange frameworks is one approach to theoretically understanding the unique governance context of the PRD region. The intricacy of the IAD framework in cataloging actors, interactions, interests and environmental characteristics blends productively with the fragmented and multi-scalar governance setting captured by the Jensen and Lange framework. In a 2014 white paper,[49] the Chinese government underscored its view of the Hong Kong-China relationship as one country with two systems. As such, a framework accounting for transboundary water governance should accommodate two systems within a single analytical sphere, as the transboundary governance setting cannot be analysed as a single unit with uniform characteristics. This theoretical proposal is an attempt to overcome that challenge and aims to provide an analytical model for similarly situated contexts.

Despite tensions between China's central government and politically active groups in Hong Kong, transboundary collaboration on practical issues like water and transportation is functional and largely apolitical. However, recent support for self-sufficiency, as articulated by Hong Kong's TWM strategy, may imply that the city's confidence in the durability of current supply arrangements is weakening. Investment in alternative supply, more than 40 years prior to the contract expiration date, could be seen either as an interest in developing technologies with a lower per-unit cost than that of imports, or as growing uncertainty about whether China is able and willing to honor the contract in the long-run.

This chapter has explored the potential for intergovernmental collaboration to enhance resilience against challenges to the management of water. Future studies should adopt a longitudinal perspective in exploring the relationship among resilience to water threats, institutional arrangements and

---

[49] SCIO. (2014). *'One Country Two Systems' in Hong Kong SAR Practice*. Beijing: State Council Information Office of the People's Republic of China. http://news.xinhuanet.com/gangao/2014-06/10/c_1111067166.htm.

policies, and intergovernmental collaboration, with a focus on the feasibility of a region-wide water governance regime that harmonizes the approaches of individual agencies and governance units. A broader research agenda should focus on the important but largely overlooked the link between the structure of local governance systems and the efficacy of regional or international agreements in managing environmental threats. Faith in the stability of imported supply, as supported by the presence of long-term contracts, has arguably bred a sense of moral hazard in Hong Kong's water policy. Remnants of colonial-era water strategies, as they survive in current policy, deserve a fundamental revisitation, and the TWM strategy includes language that signals the need for such change.

A looming challenge for both the practice and study of water management in the PRD is the evolving nature of the relationship between Hong Kong and mainland China. The border is progressively dissolving and integration gathering pace in advance of the 2047 date at which the city's 'handover' to China is complete. From a practical point of view, Hong Kong's continued operational collaboration with the mainland on water issues helps both sides cultivate a deeper level of trust, reciprocity and mutual interest—factors that might be described as 'administrative capital' that strengthens regional capacity to manage threats to water supply and numerous other environmental, economic and social matters. The changing dynamics around transboundary issues, including gradual integration, also presents scholars with an opportunity to observe a natural experiment in progress—particularly useful for longitudinal studies. Until full integration occurs, however, the transboundary setting is likely to continue to be politically sensitive and the optics of continued collaboration with the mainland will need to be managed by the Hong Kong government in an setting where scepticism about integration persists.[50] Water management provides an instructive context for this dynamic as its urgency can be understood by policymakers and the public alike. The framework applied in this chapter represents a useful path for understanding these important but complex issues.

[50]Hartley, K., Tortajada, C., & Biswas, A. K. (2018). Political Dynamics and Water Supply in Hong Kong. *Environmental Development, 27*, 107–117.

Mapping the Water Disputes in India: Nature, Issues and Emerging Trends

Ruchi Shree

INTRODUCTION: UNDERSTANDING WATER DISPUTES

As per a Buddhist fable,[1] two clans namely the Koliyas and the Sakyas lived on the banks of River Rohin are said to have fought over use of the river water. It was in the time of drought that each clan suspected the other and since Buddha mediated, the fight stopped. In the present context, in Maharashtra, many villages in Solapur and Nashik districts are accusing their respective upstreams for not releasing waters,[2] but there seems to be no possibility of mediation. So, one may say that the history of water disputes in India has a certain continuity. It also suggests that water as a natural resource being so critical for survival of living beings involves

---

[1]Silva, S. N. (2009, May 10). Conflict Resolution: How Lord Buddha's Way Offers an Answer. *The Sunday Times*. http://www.sundaytimes.lk/090510/News/sundaytimesnews_29.html. Accessed November 24, 2018.

[2]Biswas, P., & Iyer, K. (2018, November 22). East and West: Maharashtra Districts Fight Water Wars. *The Indian Express*. https://indianexpress.com/article/india/east-and-west-maharashtra-districts-fight-water-wars-drought-5458259/. Accessed November 24, 2018.

R. Shree (✉)
Department of Political Science, Janki Devi Memorial College,
University of Delhi, New Delhi, India

© The Author(s) 2020
A. Ranjan (ed.), *Water Issues in Himalayan South Asia*,
https://doi.org/10.1007/978-981-32-9614-5_5

103

104 R. SHREE

politics which is often about the power dynamics involving the state and non-state actors.

In recent times, the emergence of disciplines viz. hydro-diplomacy and water governance in India indicate how critical water is for the development of our country and the need to engage with its different dimensions. These disciplines are interdisciplinary in nature but politics remains the central issue and no wonder that many new terms viz. water security, water stress, water wars, water mining, water footprint, water automatic teller machines (ATMs), water markets and several others are slowly becoming part of the mainstream. Linton seems to have rightly made a provocative statement 'water is what we make of it'[3] in the context his tracing of the 'social construction' of the hydrologic cycle concept. In this chapter on water disputes in the Indian context, I intend to argue that water as a theme for research sans the binary between science and social science. For instance, the scientific understanding of water as H_2O forever changed our perception of water. In this paper, I follow the theoretical framework of Ivan Illich[4] and Ashis Nandy[5] to argue that modernity in general and science and state as an actor[6] as its byproducts often shape the nature of disputes of different kinds over water. Be it at the level of perception of water as to whether it is commons or commodity, or the issue of ownership viz. state is the owner or a trustee, etc. remain at the forefront and difficult to resolve.

Given the availability of vast literature on water, I would like to state that my focus remains on the domestic aspect of water disputes in India and an interesting work which has partly attempted to do the same is *Water Conflicts in India: A Million Revolts in the Making.*[7] However, since the book

---

[3] Linton, J. (2010). *What Is Water? The History of a Modern Abstraction* (p. 1). Vancouver and Toronto: UBC Press.

[4] Illich, I. (1986). *H₂O and the Waters of Forgetfulness.* New York: Marion Boyars.

[5] Nandy, A. (2003). *The Romance of the State: And the Fate of Dissent in the Tropics.* New Delhi: Oxford University Press (OUP).

[6] State in the role of planner and designer. For details, please see Shree, R. (2018). *Rivers as Commons: Reality or Myth?* as guest blog on SANDRP website. https://sandrp.in/2018/03/14/rivsers-as-commons-reality-or-myth/. Accessed March 18, 2018.

[7] Joy, K. J., et al. (2008). *Water Conflicts in India: A Million Revolts in the Making.* New Delhi: Routledge. This book has summarized 63 case studies from all over India by categorizing them into eight categories—(i) Contending Water Uses (ii) Equity, Access and Allocations (iii) Conflicts Around Water Quality (iv) Sand Mining (v) Micro-level Conflicts (vi) Dams and Displacement (vii) Tran boundary Water Conflicts (viii) Privatisation.

MAPPING THE WATER DISPUTES IN INDIA ... 105

was an initiative of a nongovernmental organization (NGO) named Society for Promoting Participative Ecosystem Management (SOPPECOM),[8] it has merely focused on classification and documentation of the ongoing conflicts in India. Similarly, two other recent books viz. an edited volume by Ramaswamy Iyer titled *Living Rivers, Dying Rivers* and Victor Mallet's *River of Life, River of Death: The Ganges and India's Future* have mainly dealt with the issue of river water but I intend to take up an interdisciplinary approach to delve into the complexities of politics of water. The chapter begins with brief sections on methodology and theoretical framework and then moves on to the laws, policies and institutional framework around water in India. It is followed by the typology of the major disputes on water in India and the critical analysis of some of the emerging trends.

METHODOLOGY

The multidimensional 'politics of water' in India got further complicated with the adoption of New Economic Policy (NEP) in 1991 which led to privatization in the water sector. It also intensified the role of civil society organizations (CSOs) in protesting against privatization and in strengthening the demand for 'right to water'. The non-state actors ranging from the international organizations viz. World Bank and International Monetary Fund (IMF) to the Multi-National Corporations (MNCs) and non-governmental organizations (NGOs) play an equally critical role as the State. Another instance is the opposition to National Water Policy (NWP 2012) guided by the neo-liberal philosophy. In this paper, the original document of NWP (2012) has been used as a primary source to highlight the position of Indian state viz-a-viz water as a natural resource. I have used the method of Critical discourse analysis[9] (CDA) to engage with this document due to two reasons—first, it generated much debate among the civil

---

[8] SOPPECOM is an NGO based in Pune (Maharashtra) and works on issues such as watershed management, participatory development, etc.

[9] Critical discourse analysis is a relatively new method in social sciences for doing analytical research. Some of its tenets could be traced in the critical theory of the Frankfurt school in the before the Second World War. CDA gained popularity in 1990s and Van Dijk writes that the CDA is a type of discourse analytical research which primarily aims to study the way social power abuse, dominance, and inequality are enacted, reproduced, and resisted by text and talk in the social and political context. T. A. Van Dijk. (n.d.). *Chapter 18: 'Critical Discourse Analysis'*, 352–371. http://www.discourses.org/OldArticles/Critical%20discourse%20analysis.pdf. Accessed June 24, 2013. To Dijk, through such dissident research, critical

society activists and second, it also reflects the prominence of development discourse and commodity perspective shaping the nature of water disputes in Indian context.

The complexity of politics around water and the disputes over its allocation as a natural resource have been analysed at many levels. At one level, it deals with actors (local and global, state and non-state, etc.), on the other it maps the issues (groundwater depletion, privatization, inter-sectoral allocation, etc.) as well as laws, policies and institutional framework on water. Further, the chapter critically engages with the nature and emerging trends of disputes over water such as convergence of macro and micro aspects, manifest and latent conflicts, politics of perspectives/world views, dominance of development discourse to name a few. The chapter focuses upon the nuances of the domestic aspects of the disputes. Taking up an interdisciplinary approach, it attempts to situate the water disputes in Indian context within the ongoing global debates on water.

Along with draft NWP, 2012, the chapter has engaged with the alternative NWP proposed by Ramaswamy Iyer. His numerous writings on water also helped me in strengthening the argument of politics if perspectives on water. Among other works on water, the author was benefitted by several interactions with Anupam Mishra,[10] Philippe Cullet[11] and Amita Baviskar[12] over a period of time. Due to my research and writings on different aspects of water for more than a decade, some of the ideas and materials have been unavoidably repeated in this chapter.[13]

---

discourse analysts take explicit position, and thus want to understand, expose, and ultimately resist social inequality.

[10] Anupam Mishra (1948–2016) was a noted Gandhian and environmentalist and his writings on water focused on traditional ways of water harvesting is known worldwide. For details see Shree, R. (2018, February). No Pretense, No Armour: Anupam Mishra and His Great Contributions. *The New Leam, 4*(33). http://thenewleam.com/2018/03/no-pretense-no-armour-anupam-mishra-great-contributions/. One may also see his TED talk https://www.ted.com/talks/anupam_mishra_the_ancient_ingenuity_of_water_harvesting?language=en.

[11] Philippe Cullet is professor of Law at School of Oriental and African Studies (SOAS, London) and written on water and sanitation laws and policies in India.

[12] Amita Baviskar is professor of sociology and has immensely worked on politics of water, especially from the vantage point of social movements and cultural politics of water as a natural resource. Her books *In the Belly of the River: Tribal Conflicts over Development in the Narmada Valley* (OUP, 1995) and (ed.) *Waterscapes: The Cultural Politics of a Natural Resource* (Permanent Black, 2007).

[13] Some of them are Shree, R. (Forthcoming). *Politics of Water as Natural Resource: Prospects of Commons Perspective*, NMML Occasional Paper; 'Money Can't Buy the Elixir of

THEORETICAL FRAMEWORK

Till the 1970s, studies on water were mainly of two types—first, engineering and hydrology approaches and second, anthropological literature on water. From the 1970s onwards, 'management' of water resources was triggered by problems in the functioning of the infrastructure and the establishment of institutes for management of water started in South Asia. In the late 1980s, the ideological tide shifted to neo-liberal focus on property rights and market mechanisms as tools for modernizing water management. In the 1990s and 2000 onwards, the critique of the underlying concept of development primarily came from social movements, especially the opposition to large dams, etc. There was a growth of 'public sociologies' of water resources management which was closely associated with civil society activism, reflective and critical of the assumptions behind state-directed water resource development.[14] Peter Mollinga argues that the writings on water in the last ten–fifteen years have expanded in 'quantitative' as well as 'qualitative' aspects.[15] He has even developed an approach to comparative research on water resources management. Today, due to disputes of different kinds, water resources development has become a complex and multidimensional phenomenon. Since, it is closely linked to societal development; it needs to have an integrated, holistic and interdisciplinary approach.

In 1970s water was recognized as a natural resource under stress and this attracted many scholars and policymakers to undertake research on this issue. Here I would like to mention that 1973 onwards, every third year World Water Congress is held. In the United Nations (UN) Water Conference at Mar Del Plata in Argentina (1977), its final declaration stressed that 'all people, whatever their stage of development and their social and economic conditions, have the right to have safe access to drinking water in quantities of a quality equal to their basic needs'.[16] The UN

Life' in The Pioneer (Op-Ed), April 29, 2016. http://www.dailypioneer.com/columnists/oped/money-cant-buy-elixir-of-life.html. 'Are We Serious About Our Rivers' in The Pioneer (Op-Ed), March 30, 2017. http://www.dailypioneer.com/columnists/oped/are-we-serious-about-our-rivers.html. Rivers as Commons: Reality or Myth? as guest blog on SANDRP website. https://sandrp.in/2018/03/14/rivsers-as-commons-reality-or-myth/.

[14] Dutt, K. L., & Wasson, R. J. (Eds.). (2008). *Water First: Issues and Challenges for Nations and Communities in South Asia*. New Delhi: Sage.

[15] Mollinga, P. P., & Gondhalekar, D. (2012). *Theorising Structured Diversity: An Approach to Comparative Research on Water Resources Management* (ICCWaDS Working Paper No. 1).

[16] Conca, K. (2006). *Governing Water: Contentious Transnational Politics and Global Institution Building*. Cambridge: The MIT Press.

108   R. SHREE

declared 1980–1990 as International Drinking Water Supply and Sanitation Decade. In 1987, Brundtland Commission released its report 'Our Common Future'. However, it has very little to say on water. In 1992, the Dublin Principles declared at the International Conference on Water and the Environment reflects a paradigm shift with its emphasis on water should be considered as an economic good. The model of the Washington Consensus has led to governments and international institutions advocating the privatization and commodification of water.

The chapter attempts to theorize the water disputes in the post-independence Indian context. As mentioned in the methodology part, it aims at mapping the complexity of water disputes in Indian context at an ideational level as well as at the level of contemporary issues and emerging trends. While stressing on the need to engage with the same in a holistic manner, the chapter explores the role of state and non-state actors in shaping the 'politics of water' around multiple issues viz. laws, policies, etc. The paper is based on two lines of arguments—first, the politics of water is essentially 'politics of perspectives'. At the moment, three conflicting perspectives are operating in India, namely, 'commons', 'commodity' and 'entitlement' perspective. Each of them as a world view has a distinctive meaning and series of implications. Their reflection could be seen in the language used in the policies to protests/movements to the court judgements. The second argument is that the 'development discourse' has shaped the politics of water in India in a significant way. To illustrate it, the chapter will touch upon four issues namely—inter-state river water disputes, anti-dam movements, inter-sectoral water allocation and privatization in water sector.

WATER IN INDIAN CONTEXT: LAWS, POLICIES AND LANDMARK COURT JUDGEMENTS

Water as a natural resource and a life giving substance is also one of the most contested resources. The contestation is rooted in the difference of world-views on water to different types of institutional arrangements to regulate and distribute it. As mentioned earlier, water governance is complex in the federal set up of India. At the Centre, Ministry of Water Resources, River Development and Ganga Rejuvenation[17] is the main body to take up the

---

[17] Under the Modi government at the Centre, the Ministry of Water Resources (MoWR) has been renamed as Ministry of Water Resources, River Development and Ganga Rejuvenation as this govt. has taken up the cause of cleaning the River Ganga to a new level through its Namami Gange Yojana.

MAPPING THE WATER DISPUTES IN INDIA ... 109

administrative tasks and the National Water Policies have been formulated by this ministry. But other ministries viz. Ministry of Agriculture, Ministry of Environment, etc. also get involved indirectly when it comes to issues such as irrigation, pollution, etc. The Central Ground Water Authority (CGWA) was created under the Environment Protection Act, 1986 'to regulate and control development and management of groundwater resources in country'.[18] To understand the nuances of politics around water, this part of the chapter engages with the provisions on water in Indian Constitution, the ramifications of federal framework, national water policies and landmark court judgements on water.

The Provisions on Water in Indian Constitution

The Indian Constitution follows the scheme introduced in the Govt. of India Act (1935), where water is a state subject. The articles and entries related to water in Indian Constitution and the federal structure of Indian state make the politics around water an interesting area of study. Though numerous instances in the Constitution may have a bearing on water but in the ambit of the paper, this section will focus on Entry 17 in the State List, Entry 56 in the Union List and Article 262.

Entry 17 in the State List states, 'Water, that is to say, water supplies, irrigation and canals, drainage and embankments, water storage and water power subject to the provisions of Entry 56 of List I'. The Constitution states water as a State subject but at the same time certain restrictions are also laid down regarding the use of interstate water. The provisions of Entry 56 in the Union List mentions, 'Regulation and development of inter-State rivers and river valleys to the extent to which such regulation and development under the control of the Union is declared by Parliament by Law to be expedient in the public interest'.[19] Article 262 has bearing upon the Centre–State relations regarding water as it states the process of adjudication of inter-state water disputes (ISWD). The Inter-State Water Disputes Act (1956) was regulated as per this provision.

---

[18] For details, see http://cgwb.gov.in/aboutcgwa.html.

[19] Indian Constitution, Schedule 7, List 1, Entry 56.

110   R. SHREE

Ramaswamy Iyer writes 'in the absence of a specific mention, it has been generally assumed that the reference to "water" in the Constitution includes groundwater'. Even in the Fundamental Rights Section, 'the right to life has been held by judicial interpretation to include the right to water as a life-sustaining resource, and here too, "water" has been taken to include groundwater'.[20] The issue of water resources in terms of surface water and groundwater does not have clarity in the Indian Constitution and thus it often leads to ambiguity in laws and policies.

National Water Policies[21]

It is interesting to note that for forty years after independence, no national policy was made on water. First NWP came up in 1987 and talked about conjunctive use of surface water and groundwater. It was revised in 2002 and as a result second NWP was formulated by the Ministry of Water Resources. Uttam Kumar Sinha argues 'while the water policy lays strong emphasis on the existential, economic and ecological values of water, less attention is given to principles and objectives to achieve sustainable and equitable use of water'.[22] The NWP, 2002 stipulated that the state governments should formulate State Water Policies in accordance with the NWP.[23]

NWP, 2002 was further edited as draft NWP, 2012 which clearly marks a step ahead of the previous water policy. The gap of ten years between the second NWP and the draft of the third NWP seems to have strengthened Indian State's commitment to neo-liberal policies. It starts with a few often quoted data on water scenario in India underlining the rising situation of acute water scarcity i.e. 'India has more than 17 percent of the world's population, but has only 4% of world's renewable water resources with 2.6% of the world's land area'. In the preamble, the draft outlines its objective

---

[20] Iyer, R. (2003). *Water: Perspectives, Issues, Concerns* (p. 101). New Delhi: Sage.

[21] This part of the chapter heavily relies on one of the chapters of my unpublished Ph.D. thesis titled 'Politics of Water as Natural Resource: Study of Two Movements (Plachimada and Tarun Bharat Sangh)' (2014) at Centre for Political Studies, Jawaharlal Nehru University. Some other parts of the chapter also draw partly from the same.

[22] Sinha, U. K. (2014). The Strategic Politics of Water in South Asia. In J. Miklian & A. Kolas (Eds.), *India's Human Security: Lost Debates, Forgotten People, Intractable Challenges*. New Delhi: Routledge.

[23] Ibid., Uttam Kumar Sinha.

'to take cognizance of the existing situation and to propose a framework for the creation of an overarching system of laws and institutions and for a plan of action with a unified national perspective'.

The draft underlines the issues of concern like rising water stress due to population growth, urbanization, changing lifestyles; wide temporal and spatial variation in availability of water; climate change; unequal access to drinking water likely to cause social unrest; groundwater as individual property being exploited inequitably; inter-state, inter-regional disputes over sharing water; growing pollution of water sources; low public awareness of rising water scarcity, etc. All these concerns are well defined but at the same time, if one reads this draft between the lines, its overemphasis on the fact of 'scarcity' is very evident. It becomes important to bring Lyla Mehta's argument[24] how 'scarcity' itself is both 'real' and 'constructed'. She says that many a times the situation of 'water scarcity' is manufactured through political and policy processes and it is experienced by the combination of sociopolitical, discursive and institutional factors. Mehta also makes a crucial point that the way state discourses and programmes essentialize scarcity as a natural phenomenon, it enables the political legitimization of large dams. This process also tends to marginalize the local knowledge systems.

The draft policy also states that water policies must be based on principles of equity and justice in the allocation of water. But at the same time, in the preamble, it also states that there is a need to recognize, 'water, over and above the pre-emptive need for safe drinking water and sanitation, should be treated as an economic good so as to promote its conservation and efficient use.' In the second section on 'water framework law', the draft has mentioned, '…it is recognized that States have the right to frame suitable policies, laws and regulations on water; there is a felt need to evolve a broad over-arching national legal framework of general principles on water'. These statements from the draft very well reflect the neo-liberal agenda of Indian state where it is overemphasizing the shift in the very status of water from a 'social good' to 'economic good'. This shift also reflects our policies following the model in the West, everything has to be sellable. The gradual transformation and notion of commodity for even water seems to be slowly becoming acceptable among common people.

---

[24] Mehta, L. (2013). Contexts and Constructions of Water Scarcity. *EPW, 38*(48). She has focused on the 'water-scarce' Kutch and its relationship with the controversial Sardar Sarovar Project.

Another statement that '...water needs to be managed as a community resource held, by the state, under public trust doctrine to achieve food security, livelihood, and equitable and sustainable development for all' also seems very problematic. Here, the notion of 'community resource' and 'held by the state' appears as contradictory to each-other. India has a rich tradition of water sources like wells and ponds managed by the temples,[25] how will such traditions be managed? Similarly, the proposal in the draft that 'the Indian Easements Act, 1882[26] may have to be modified accordingly in as much as it appears to give proprietary rights to a land owner on groundwater under his/her land' could have both its positive and negative consequences. It is more or less very clear that the state wants to have its control upon not only surface water but also the groundwater.

Landmark Court Judgements Related to Water

There are many court cases related to water in India. However, I intend to selectively mention about those cases which have invoked the issue of water as a resource. In the last two decades, many cases have come up where the Supreme Court has tried to make a scope for 'right to water' as part of 'right to life' under Article 21. As such, 'right to water' is not defined in Indian Constitution but human rights movement gaining strength all over the world, even in India, the apex court has played an important role in addressing the related issues like the right to health, education, food, water, etc. One may club the court cases and important judgements related to water in two categories—first, where right to water has been defined as part of right to life; and, second, rights against pollution of water sources

---

[25] An important study has been done by David Mosse (2003) in the context of South India.

[26] Under the Indian Easement Act, 1882, Indian law recognizes the right of a riparian owner i.e. someone who owns the land adjoining a river or water stream to have unpolluted waters. Section 7 of the Act provides that every riparian owner has the right to the continued flow of waters of a natural stream in its natural condition without destruction or unreasonable pollution. There are various court cases in which the rights of the riparian have been reinforced.

The Easement Act also recognizes the customary rights of the people which are required under two rules: long usage or prescription and local custom. However, these rules were also subject to the government's right to regulate the collection, retention and distribution of water of rivers and streams flowing in natural channel.

Pant, R. (2003). *From Communities' Hands to the MNCs' BOOTs: A Case Study from India on Right to Water*. Uttaranchal, India: Ecoserve, also available at 16 Rights and Humanity, UK (Right to Water Project). http://www.righttowater.org.uk/pdfs/india_cs.pdf.

MAPPING THE WATER DISPUTES IN INDIA ... 113

(like ponds, rivers, etc.). But, since two of them are closely linked to each other, we will do it as analysing some of the important judgements by the apex court of India.

While upholding the Indian government's decision to construct over 3000 dams on river Narmada, the Supreme Court stated that 'water is the basic need for the survival of human beings and is part of the right to life and human rights as enshrined in Article 21 of the Constitution of India...and the right to healthy environment and to sustainable development are fundamental human rights implicit in the right to life'.[27] In some other cases[28] also, the Supreme Court has repeatedly reaffirmed the proximity between public access to natural resources, including water, the right to a healthy environment and the right to life under Article 21 of the Indian Constitution.

The Supreme Court has played quite an active role in the context of defining the state's duty not to pollute, to mandate the cleaning up of water sources and the coastline by the polluter and restitution of soil and groundwater.[29] Many a times, the court has also applied the 'precautionary principle' to prevent the potential pollution of drinking water sources consequent upon the setting up of industries.[30] It is very important to understand that the role of court judgements becomes very crucial in defining the role of the state. The judiciary also issues notices to the government in case of its non-compliance to the principles underlined in the Constitution.

[27] Narmada Bachao Andolan v. Union of India AIR 2000 SC 3751; 248(2000) 10 SCC 664.

[28] Subhash Kumar v. State of Bihar AIR 1991 SC 420. In this case, the person had filed the case under Public Interest Litigation (PIL) to prevent the pollution of the Bokaro river water from the sludge/slurry discharged by the Tata Iron and Steel Company (TISCO). But later on, the court found the petitioner to have made a false allegation due to a personal grudge against the company. However, the recognition of 'right to pollution free water' makes this case very important. For details, one may see Ruchi Pant's article (2003: 14–15).

[29] Narain, V. (2009). Water as a Fundamental Right: A Perspective from India. *Vermont Law Review, 34*, 1–9.

[30] Muralidhar, S. (2006). The Right to Water: An Overview of the Indian Legal Regime. In E. Riedel & P. Rothen (Eds.), *The Human Right to Water* (pp. 65–81). http://www.ielrc. org/content/a0604.pdf. Accessed March 9, 2011. He has mentioned about various cases like M.C. Mehta v. Union of India AIR 1988 SC 1037; M.C. Mehta v. Kamal Nath (1997) 1 SCC 388; S. Jagannath v. Union of India (1997) 2 SCC 87; A.P. Pollution Control Board v. Prof. M.V. Naydu (1999) 2 SCC 718; A.P. Pollution Control Board (II) v. Prof. M.V. Nayudu (2001) 2 SCC 62.

TYPOLOGY OF WATER DISPUTES IN INDIA

The complexity of water as a resource and the disputes over it originates from a range of diverse issues being attached to it—be it its properties or characteristics; its governance or numerous claimants with their range of ideas regarding its management. There are traditional, religious and spiritual claims to it as well as modern scientific claims for its upkeep. There is a global politics of it as well as regional, national and local ones. Sometimes what seems scientific and rational about water to some scholars while others criticize the very same on other grounds.

Inter-state River Water Disputes

The federal framework of Indian state with a tilt towards a strong Centre and the States of different sizes falls in the category of asymmetrical federalism. As mentioned earlier, the Inter-State Water Disputes Act was regulated in 1956. According to this, specific tribunals are created for addressing interstate water disputes. This Act has been used in landmark disputes concerning the Cauvery, Narmada and Krishna–Godavari rivers.[31] The Krishna–Godavari began in 1951 and a key issue was whether initial agreements about diversions from the river were justified given legal and political changes following independence.[32] Radha D'Souza's book *Interstate Disputes over Krishna Waters* suggests that 'the attrition over the Krishna river, between the states of Andhra Pradesh, Maharashtra and Karnataka, is a product of particular framings of law, science and technology'.[33] Setting the debate around the region's ecology and history, she highlights the functioning of KWDT as a legal process and its inadequacies.

Narmada Water Dispute Tribunal was set up in 1979 to resolve the issue of river water sharing among Gujarat, Maharashtra and Madhya Pradesh. It provided the framework for the construction of Sardar Sarovar Dam (SSD)

---

[31] Cullet, P., & Gupta, J. (2009). India: Evolution of Water Law and Policy. In J. W. Dellapenna & J. Gupta (Eds.), *The Evolution of the Law and Politics of Water* (pp. 159–175). Dordrecht: Springer.

[32] D'Souza, R. (2006). *Interstate Dispute over Krishna—Law, Science and Imperialism.* Hyderabad: Orient Longman.

[33] D'Souza, R. (2007, April 21–27). Water as Dispute and Conflict (Review of *Interstate Dispute over Krishna—Law, Science and Imperialism* by Radha D'Souza; *Conflict and Collective Action: The Sardar Sarovar Project in India* by Ranjit Dwivedi). *Economic and Political Weekly (EPW)*, 42(16), 1431–1432.

MAPPING THE WATER DISPUTES IN INDIA ...   115

and the protests against it eventually led to Narmada Bachao Andolan (NBA) in the 1980s. Numerous issues viz. displacement, rehabilitation, livelihood, etc. came to the forefront. We will have a detailed analysis of NBA in the section on anti-dam movements. However, it is significant to mention two points here—first, the role played by the Supreme Court[34] and second, the complications of the federal framework in the Indian context in complicating the disputes over water.

Another renowned case of inter-state water dispute is the Cauvery dispute between Karnataka and Tamil Nadu. The history of this dispute is more than a century old and it has been well researched and documented from different perspectives.[35] Although the origins of dispute over Cauvery river water sharing goes back to the pre-independence days but as per ISWD Act, 1956, the Cauvery Waters Dispute Tribunal was established by Government of India in 1990. The tribunal's decision came in 2007 but Tamil Nadu sought for intervention of the Supreme Court[36] in 2016. In 2018, the Supreme Court reduced Tamil Nadu's share and increased Karnataka's which led to violent protests.

Ever since the Central Government issued a notification in 1976 under the Punjab Reorganisation Act 1966, allocating the erstwhile Punjab's share of Ravi- Beas waters to the new states of Punjab and Haryana, with a small amount to Delhi, the dispute over water is just not ready to resolve. While Cauvery dispute is a case of upper-riparian/lower-riparian dispute, this one is a case of dispute relating to allocation of river waters as per reorganization of states. So, one may say that given the continental size of India and a complex federal structure, the disputes over sharing the river water namely ISWD have a wide range of variety.

---

[34] For details, please read Cullet, P. (Ed.). (2007). *Sardar Sarovar Dam Project: Selected Documents.* Aldershot: Ashgate.

[35] Ibid., Iyer, R. (2003); Iyer, R. (2007). *Towards Water Wisdom: Limits, Justice, Harmony.* New Delhi: Sage; Guhan, S. (1993). *The Cauvery River Dispute: Towards Conciliation.* Chennai: Frontline Publications.

[36] Anonymous, HT Correspondent. (2018, February 16). From 1924–2018: Twists and Turns in the Cauvery Water Dispute. *Hindustan Times.* https://www.hindustantimes.com/india-news/from-1924-to-2018-the-twists-and-turns-in-the-cauvery-water-dispute-case/story-d5PuU5mSuUZot2NNlI7ZEJ.html. Accessed November 28, 2018.

Inter-sectoral Water Allocation

In recent past, inter-sectoral water allocation has emerged as a major issue in India. Three major sectors for allocation of water are agriculture, industry and domestic use. The Central Water Commission has recorded that irrigation takes around 80% of India's water. Agriculture sector being the largest groundwater user is now being pushed for drip irrigation for efficient use of water. The planners are also debating the need for radical transformation in the cropping pattern so that we move away from water-intensive crops such as wheat and sugarcane to pulses and millets.[37]

Some industries, especially distillery, sugar and textile consume, pollute and discharge large amounts of wastewater into the rivers. These effluents are generally toxic, poisonous and non-degradable. The same sources of water viz. rivers and lakes are often contested as per their use and misuse. For example, a lake or river could be part of development project for beautification and could also be the dumping ground for sewage. A study[38] of this kind has been done in the Rajsamand district of Rajasthan to illustrate the gaps in existing law and policy framework and multiple issues that need to be addressed on water. Many states of India are grappling with problems of water management (floods, drought, groundwater depletion, etc.) and thus trying to devise ways for efficient use of water.

Anti-dam Movements

In the 1950s and 1960s, big dams symbolized the progress of humanity and celebrated them as human control over unpredictable forces of nature. Even, in India, our first prime minister Pandit Nehru considered the large dams as 'new temples of India'. However, this notion was severely challenged by the social activists who challenged the construction of large dams in the name of displacement, rehabilitation, destruction of ecosystem, etc. This part of the chapter will focus on the anti-dam movements in India with a major focus on movement against dam on river Narmada in 1980s. However, the protests and movements by the local people as well as the

---

[37] Ibid., Shah, Mihir (2018).

[38] Cullet, P., Bhullar, L., & Koonan, S. (2015). Inter-Sectoral Water Allocation and Conflicts: Perspectives from Rajasthan. *EPW, L*(34), 61–69.

civil society activists against the dams has a wide range of diversity.[39] The geographical area of these struggles covers North, South and North-East India.[40]

The Narmada River, India's largest west-flowing river passes through three states—originating in Madhya Pradesh then reaching Maharashtra and finally emptying itself in Gulf of Khambhat (Arabian Sea) in Gujarat. In the beginning, the problem started with sharing of river water between these three states. Government of India constituted Narmada Water Disputes Tribunal to solve this problem in 1969. After ten years, the Tribunal gave its decision in 1979. As per the Tribunal's decision, 30 large dams, 135 medium dams and 3000 small dams were to be constructed. It also granted the approval to raise the height of SSD. However, later on, this decision was severely challenged by the social activists and NBA began a major movement against the large dams.

The construction of dams to harness the irrigation and hydroelectric potential of the Narmada River represents India's most ambitious development project till date. At the same time, probably it is also the most controversial development project in India. Narmada Anti-dam Agitation popularly known as NBA is one of the most famous environmental movements in India.[41] The anti-dam activists of NBA consider the big dams in general and SSP in particular as 'unmitigated environmental, social and economic disasters'.[42] Worldwide, this movement has almost gained a cult status among the movements against large dams.

Some other renowned movements against dams in India are in Kerala and Himachal Pradesh in recently in different states of North-East.[43] Most of the projects of large dams are for the production of electricity. It has led to a divide between the state and civil society. Very often, the

---

[39] For details, kindly refer to Nayak, A. K. (2010). Big Dams and Protests in India: A Study of Hirakud Dam. *EPW, XLV*(2), 69–73.

[40] Bhatacharjee, J. (2013). Dams and Environmental Movements: The Case for India's North-East. *International Journal of Scientific and Research Publication, 13*(11), 1–11.

[41] For a brief history of the struggle famous as NBA, see Sangvai, S. (2000). *The River and Life: People's Struggle in the Narmada Valley*. Mumbai: Earth Care Books.

[42] Paranjape, S., & Joy, K. J. (2006, February 18). Alternative Restructuring of the Sardar Sarovar: Breaking the Deadlock. *EPW*, 601–602.

[43] Mimi, R. (2010, December 10). *Anti-dam Protests Get Louder in North-East*. https://www.internationalrivers.org/resources/anti-dam-protests-get-louder-in-northeast-india-1689. Accessed November 25, 2018.

118 R. SHREE

social movements against the large dams are also seen as anti-development and anti-state. However, in 2016 one of the reports by Ministry of Water Resources accepted that despite spending 4 lakh crores in building large and medium dams across our rivers, we have not been able to achieve the goal of required water storage.[44] Millions of people have been displaced and the amount of environmental destruction is beyond imagination.

Protests Against Water Privatization

The story of privatization in water sector in India began in the post-New Economic Policy period. First remarkable case was privatization of 23.6 km stretch of Sheonath River was given to a private company named Radius Water Limited (RWL) in 1998 by the Madhya Pradesh State Government. Later, in 2001, the Build-Own-Operate-Transfer (BOOT) became functional under the Chhattisgarh government.[45] The protests against it by the local people and the CSOs led to cancellation of this project. Later, in Plachimada (Kerala), the anti-Coca-Cola protests[46] also led to closure of the plant. Overall, the attempts of privatization in water sector has been a failure[47] in India due.

There is a close link between anti water-privatization movements and rising demands for 'right to water'. While resisting privatization these movements have also started their demand for 'water as human right'. From NAPM to demands at Plachimada (Kerala) and Kaladera (Rajasthan),[48] everywhere the rights discourse has become so central to the debate. Another instance is the formation of 'Campaign Against Water Privatisation' in Karnataka where 40 NGOs, slum dwellers' groups, citizen's initiatives, dalit groups, etc. came together to challenge the attempts of

---

[44] Mihir Shah in an interview titled 'The Answer to Our Water Crises Lies in the Democratization of Water Resources', *Geography and You*, Vol. 18, Issue 2, No. 113 (2018), pp. 22–24.

[45] After the division of the state of Madhya Pradesh, that part of Sheonath River fell into Chhattisgarh. For details, see Das, B., & Pangare, G. (2006, February 18). In Chhattisgarh, a River Becomes Private Property. *Economic and Political Weekly*, 611–612.

[46] For details, see Shree, R. *Plachimada Against Coke: People's Struggle for Water* (published by SOPPECOM on the website of India Water Portal in 2012). http://www.conflicts.indiawaterportal.org/node/147.

[47] Pranjape, M. (2016, October 17). *Privatisating India's Water Is a Bad Idea.* https://thewire.in/politics/water-privatisation. Accessed November 25, 2018.

[48] Anti coca-cola protests by the local people and also considered as part of new social movements (NSMs).

privatization.[49] Similarly, the attempt to privatize the Sonia Vihar Plant in Delhi was challenged by the NGOs like *Navdanya* by forming 'Citizens' Alliance for Water Democracy'. Vandana Shiva's *Navdanya* along with various other NGOs like *Ganga Raksha Samiti, Ganga Mukti Andolan,* etc. has also been very active in 'Save the Ganga Movement'.[50] On similar lines, even 'Save the Yamuna Campaign' has also gained momentum in the last few years.

The non-state actors working on various issues related to water also play a key role in influencing the policymaking on water at national as well as international level. Many of these organizations are part of the International River Network (IRN), an international NGO. Its international advisory board includes Shripad Dharmadhikary of Manthan and Himanshu Thakkar of SANDRP. Other renowned water activists like Patrick McCully and Dipak Gyawali are its members of the United States Advisory Board and South Asia Advisory Board, respectively. Both of them have extensively worked upon water issues in the Indian context. So, the lobby of non-state actors around water has its national as well as global appeal in promoting the anti-privatization campaign as well as strengthening the 'right to water' discourse. However, probing deep into the issue makes it very clear that most of the CSOs approach the issue with their different ideological leanings.

CRITICAL ANALYSIS OF THE WATER DISPUTES: EMERGING TRENDS

At present, India is the largest consumer of groundwater in the world as per the reports of World Bank and reiterated by many other organizations. The major dispute on water is about the sectoral division of water. New principles viz. Water User Associaltions (WUAs), water tariff, etc. are being adopted in different parts of the country. Mihir Shah, noted economist and water policy expert argues that 'India suffers from hydro-schizophrenia that is lacking a synergy between the adopted approaches to manage and conserve water'[51] and suggests that water democratization is the answer to

---

[49] Dwivedi and others (2006: 35).

[50] Shiva, V. (2012). *Making Peace with the Earth: Beyond Resource, Land and Food Wars* (pp. 96–97). New Delhi: Woman Unlimited, An Associate of Kali for Women.

[51] Ibid., Shah (2018: 23).

water crises in India. In this backdrop, present part of chapter elaborates upon some of the contemporary trends of water disputes in India.

Politics of Perspectives

As mentioned earlier, the roots of water disputes in India lie in the politics of perspectives. Three major perspectives namely commons, commodity and entitlement seem to shape the ongoing debates and discussions as well as actions on water. The actors viz. state and non-state are informed by these perspectives and it gets reflected in the texts or documents produced by them. For example, Indian state as an actor through its draft NWP, 2012 reflects the gradual shift towards water a socio-economic good in the light of NEP. The use of terms viz. water audit and water tariff suggest the same. On the other hand, CSOs take up water as a social good and intend to assert the 'rights discourse' and Ramaswamy Iyer's alternative NWP[52] could be seen in this light. While the government is focusing on the management aspect, Iyer argues for harmonious and equitable use of water.

It is important to note that the Supreme Court has also recognized water as a community source to be held by the State in public trust in recognition of its duty to respect the principle of inter-generational equity. In *M.C. Mehta v. Kamal Nath*[53] the Court declared that 'our legal system – based on English common law – includes the public trust doctrine as part of its jurisprudence. The State is the trustee of all natural resources, which are by nature meant for public use and enjoyment. The public at large is the beneficiary of the seashore, running waters, air, forests and ecologically fragile lands. The State as a trustee is under a legal duty to protect the natural resources. These resources meant for public use cannot be converted into private ownership'.[54] This case clearly mentions the role of state as trustee. However, many times, the role of the state itself becomes doubtful. For example, the role of the state in privatization of Sheonath River in Chhattisgarh under the BOOT system.

Recently, the Supreme Court ordered directing the time bound provision of safe drinking water to 18 long-suffering residential areas around the

---

[52] Iyer, R. (2011, June 25). National Water Policy: An Alternative Draft for Consideration. *EPW, 46*(26–27).

[53] *M.C. Mehta v. Kamal Nath* AIR 2000 SC 1997 34.

[54] Ibid., Muralidhar (2006).

MAPPING THE WATER DISPUTES IN INDIA ... 121

Bhopal gas tragedy site. The editor of *The Hindu* says this order throws light on two issues—'the slow pace at which even the highest judicial intervention produces results in India and the persisting pollution threat to the health of thousands'.[55] The government has regarded this 27-year-old industrial catastrophe like other chronic problems and the residents in that area have been suffering the terrible consequences of this tragedy.

Dominance of Development Discourse

In the post-second world war period, development as a concept became so central to social sciences. The very classification of countries as developed and developing ones reflects the overarching presence of development as a discourse. Escobar writes about the making and unmaking of third world[56] and reflects on the aspirations of the developing countries to become developed ones. Politics of water also has its ramifications and as Ramaswamy Iyer writes, 'Sardar Sarovar, the Inter-linking of Rivers Project, gigantic industrial, mining and other projects, and so on, are all manifestations of a certain idea of development'.[57] The meaning of development may range from state of mind to a value in itself. In the last two decades, vast literature has come up which question the western notions of development. It has also paved the way for renewed interest in indigenous traditions of knowledge systems.

In his book, *Dams and Development: Transnational Struggles for Water and Power*, Sanjeev Khagram[58] investigates the growing struggle of local and transnational resistance against big dam projects, specifically India's Narmada Projects. Like most other countries, in India also, large dams used to be considered as the symbol of development, modernization and economic progress. The international lending agencies, such as the World Bank (WB), Food and Agriculture Organization (FAO), and Inter-American and Asian Development Banks promoted the business of dam construction. The contracts of these dams were given to the large multinational corporations (MNCs) like Siemens, Bechtel, and General Electric Corporation. So, these lending agencies and the MNCs became the targets of these movements.

---

[55] *The Hindu.* (2012, May 11). Water, and Justice, for Bhopal.

[56] Escobar, A. (1995). *Encountering Development: The Making and Unmaking of the Third World.*

[57] Ibid., Iyer, R. (2007); Ibid., p. 130.

[58] Ibid., Khagram, Sanjeev (2000).

Organized opposition to dam building began in India with 'progressive domestic institutionalization of global norms on the environment, indigenous peoples, and human rights denaturalizing big dams',[59] and it was simultaneously accompanied by the mobilization of grassroots social groups in the early 1970s. Prior to the formation of NBA, there were groups such as Gujarat-based Arch-Vahini (Action Research in Community Health and Development) and Narmada Asargrastha Samiti (Committee for people affected by the Narmada dam), Madhya Pradesh-based Narmada Ghati Nav Nirman Samiti (Committee for a new life in the Narmada Valley) and Maharashtra-based Narmada Dharangrastha Samiti (Committee for Narmada dam-affected people) who either believed in the need for fair rehabilitation plans for the people or who vehemently opposed dam construction despite a resettlement policy.

Convergence of the Macro and Micro

The man–nature relationship in Indian civilization has been based on holistic perspective of the universe and one may see its reflection in numerous rituals around worshipping the forms of nature across the religions, be it Sun, Moon or the rivers. Kapila Vatsyayan writes, 'any Indian is familiar with the daily rituals which serve as reminders of the concept of pure, and therefore, holy water'[60] and reminds of water as one of the five principal components of the environment (other four are earth, air, space and fire) which comprise the microcosm of the biological man and also the macrocosm of the universe. So, when it comes to water, one may see the convergence of the macro and the micro.

Another level of analysis on water disputes could also be seen as global and local. The paradigm of Integrated Water Resource Management (IWRM) as an overarching principle guides the politics of water in developing countries such as India. The projects viz. inter-linking of rivers, privatization in the name of water sector reforms, construction of large dams for the purpose of irrigation and production of electricity could be seen in this light. Ashis Nandy[61] argues that dam as an idea is rooted in large

---

[59] Ibid., Khagram (2000: 43).

[60] Vatsyayan, K. (2010). Introduction: The Ecology and Myth of Water. In *Water: Culture, Politics and Management, India International Centre.* New Delhi: Pearson.

[61] Ibid., Nandy (2003).

scale investment in nature facilitated by modern science. It is legitimized by the politicians along with bureaucrats and technocrats.

CONCLUSION

In the recent past, the ever-increasing literature on water has named it as liquid gold, transboundary resource,[62] peculiar resource[63] and several others. This chapter on mapping the water disputes in India has dealt with water as a complex resource. The complexity arises due to varied actors (state and non-state) and issues (groundwater depletion, irrigation, pollution, conservation to name a few). The chapter has also discussed how and why the politics of perspectives around water shape the disputes over water. Water as commons, commodity and entitlement are three different discourses which mould the actions and worldviews of the state and non-state actors. As mentioned in the chapter, the demand for right to water is gaining ground in India after the UN declaration in 2002[64] of the same and the Supreme Court interpreting it as part of right to life under Article 21.

The chapter also argues that the development discourse as a guiding principle paved the way for large dams and privatization in water sector but it also led to intensification in the activities by the civil society organizations. At the end, let us go back to the Buddhist fable mentioned in the beginning of the chapter. Buddha played the role of mediator and resolved the dispute between the two clans fighting over the use of river water but today despite the presence of numerous institutions, we are unable to resolve most of the disputes on water around us. It seems that modernity as a worldview (be it reasoning as a technique or modern science progressing so fast) has not equipped us enough to overcome the dilemmas. No wonder, from the awe of science and technology and thus controlling the nature through construction of large dams, we are now gradually moving towards environmental economics and river ecology to name a few.

---

[62] Ibid., Sinha, U. K. (2014).

[63] Jacob, N. (2018). A Scarcity of Management. *Geography and You*, Vol. 18, Issue 2, No. 113, pp. 6–7.

[64] UNESC. (2002). *Substantive Issues Arising in the Implementation of the International Covenant on Economic, Social and Cultural Rights* (General Comment No. 15). Geneva: United Nations Economic and Social Council. www.unchr.ch. One may also see Shree, R. (2010, December). Water as a Natural Resource: Right Versus Need Debate. *Rajagiri Journal of Social Development, 2*(1), 1–24.

Multi-stakeholder Hydropower Disputes and Its Resolutions in Nepal

Sanju Koirala, Prakash Bhattarai and Sarita Barma

INTRODUCTION

Nepal is rich in water resources and has a huge hydropower potential. Nepal has more than 6000 rivers and rivulets with a total length of about 45,000 kilometres and has 2.27% of world's water resources.[1,2] The perennial type of river run-off flowing through the steep gradient in many parts of the country makes ideal conditions for the generation of hydropower. Nepal has the theoretical potential to generate 83,000 MW of hydropower, of which 43,000 MW is estimated to be economically feasible for production.[3] While hydropower is the dominant source of electricity production,

---

[1] Hussain, I., & Giordano, M. (Eds.). (2004). *Water and Poverty Linkages: Case Studies from Nepal, Pakistan and Sri Lanka* (Project Report 1). Colombo, Sri Lanka: International Water Management Institute (IWMI).

[2] Sangroula, D. P. (2009). Hydropower Development and Its Sustainability with Respect to Sedimentation in Nepal. *Journal of the Institute of Engineering, 7*(1), 56–64.

[3] Government of Nepal. (2005). *National Water Plan*. Kathmandu: Water and Energy Commission Secretariat.

S. Koirala (✉) · P. Bhattarai · S. Barma
Center for Social Change, Kathmandu, Nepal

© The Author(s) 2020
A. Ranjan (ed.), *Water Issues in Himalayan South Asia*,
https://doi.org/10.1007/978-981-32-9614-5_6

125

it only accounts for about 2% of the total energy usage in Nepal.[4] The dominant form of energy used in Nepal is still biomass with 88% of population consuming it. By the end of 2016, Nepal was only producing 802.4 MW of hydropower.[5] The latest data on the population with access to electricity is not available however, until 2010 only 40% of the Nepali population had access to electricity.[6] For a long period of time even those who had access to electricity services suffered from 14 to 18 hours of power cut in a day during dry season. Nepal has been ranked as 137 out of 147 countries in terms of quality of electricity supply.[7]

In order to solve the problem of severe power shortage, the Government of Nepal (GoN), particularly after 2006's political changes,[8] has given a high priority to the development of hydropower projects. For this, the government has also encouraged private sector and international investors to invest in hydropower production in Nepal. The present government led by Nepal Communist Party has set a target to produce 10,000 MW of electricity in the next 10 years i.e. until 2028 for domestic consumption and foreign export. To meet this target, the government so far has given license to 172 companies to construct hydropower plants. Likewise, 36 hydropower projects have completed the feasibility study mandated by the government and applied for the license to construct projects.[9] Although the government has announced its ambitious plan for large-scale hydropower production, it is equally challenging to materialize not only because of technical reasons, but also due to political factors, divergent interests, and different perceptions among stakeholders regarding the construction of hydropower

[4]Government of Nepal. (2018). Retrieved from https://www.nepal.gov.np/NationalPortal/view-page?id=92. Accessed October 2, 2018.

[5]Asian Development Bank. (2017). *Nepal Energy Sector Assessment, Strategy and Road Map.* Kathmandu: ADB.

[6]Water and Energy Commission Secretariat. (2010). *Energy Sector Synopsis Report Nepal.* Kathmandu: Ministry of Energy.

[7]Ibid.

[8]King led government was collapsed, democracy was restored, and decade armed struggle come to an end through a peace agreement signed between the government and the Maoist rebels.

[9]Government of Nepal. (2018). *Energy, Water resources and Irrigation Sector's Current Status and Future Roadmap* (White Paper). Kathmandu: Ministry of Energy, Water Resources and Irrigation.

projects.[10] As a result, they attempt to influence the implementation process which many times foster hostile situations between the stakeholders and creates unfavourable environment for the successful implementation of the project. In some cases, such projects are trapped in a long gestation period and in some the construction works are delayed for months or years.[11] In this regard, it is pivotal to understand the factors behind arising disputes among different hydropower stakeholders and processes, methods and strategies adopted in resolving those disputes.

The central focus of this chapter is to understand the nature and characteristic of disputes in hydroelectric projects selected as cases for this study and identify similar as well as divergent disputing issues prevalent in these projects. This chapter also provides an insight into why some disputing issues are common in all hydroelectric projects and other issues not. Having such understanding can provide useful inputs to the policymakers for developing a common dispute resolution strategy to address the disputes observed in different hydroelectric projects. This chapter also attempts to compare different nature of hydropower disputes and the resolution process and strategies adopted in each project. Having such understanding provides a useful insight into the successful dispute resolution processes and strategies, which can be replicated in other context as well. In addition, this chapter also provides other useful insight into the governmental, non-governmental, and private sector initiatives for resolving hydroelectric disputes in Nepal.

This study has a few limitations as well. First, the chapter is mostly based on the secondary sources of information obtained through the review of policy documents, academic resources, newspaper articles, and web-based information posted by relevant governmental, non-governmental, and private institutions engaged in hydroelectric affairs in Nepal. Thus, it lacks field-based research and interviews with relevant individuals. Second, this study only incorporates the broader study of three contemporary hydroelectric projects, which may miss out a number of micro-level disputes associated with hydroelectric projects. Finally, because of authors' existing professional background, this chapter is written from the scholar-practitioner perspective. Thus, some of the issues highlighted in this chapter

---

[10]World Commission on Dams. (2000). *The Report of the World Commission on Dams and Development*. London and Sterling: Earthscan.

[11]Koirala, S. (2015). *Hydropower Induced Displacement in Nepal*. Dunedin, New Zealand: University of Otago.

128   S. KOIRALA ET AL.

are based on authors' own understanding of Nepal's hydropower development process and disputes observed during the processes.

This chapter is structured in following ways. In the next section, authors
provide an overview of hydropower disputes in Nepal. Authors then provide an overview of stakeholders associated with hydropower projects and
occurrence of multi-scale politics in hydropower construction. In the subsequent section, authors provide information about three hydroelectric
projects selected as cases for this study and dispute observed in each case.
This paper then provides a synchronized analysis based on the description
of cases presented in this chapter. Finally, some conclusions and recommendations are provided.

An Overview of Hydroelectric Disputes in Nepal

Nepal has more than 100 years of history of hydropower production which
starts with the construction of 500 Kilowatt Pharping hydroelectric project
in 1911 with the initiation of Rana[12] Prime Minister Chandra Shamsher
to light his palace.[13,14] During Panchayat[15] era between 1960 and 1990,
hydroelectric projects such as Kulekhani, Marshyangdi and few others were
built to meet the electricity needs of the country, particularly in city areas.
The actual proliferation of hydroelectric projects began after the restoration
of multiparty democracy in Nepal in 1990. Subsequently, Arun III, West
Seti, Upper Karnali, Upper Tamakoshi, Pancheshwar Multipurpose and
many other hydropower projects came into debates and discussions after
the restoration of multiparty democracy.

Due to authoritarian regime during Panchayat era, people hardly raised
their voice against any development projects commenced by the government. People's demand for compensation for the loss, resettlement and
benefit sharing were minimal and were often forced to accept whatever is

---

[12] Rana is a clan who ruled in Nepal for 104 years until they were made to step down through
a people's revolution in 1950.

[13] Sharma, R. H., & Awal, R. (2013). Hydropower Development in Nepal. *Renewable and
Sustainable Energy Reviews, 21*, 684–693.

[14] Koirala, S. (2015). *Hydropower Induced Displacement in Nepal.* Dunedin, New Zealand:
University of Otago.

[15] Panchayat is another autocratic/no party regime led by King which lasted for 30 years
from 1960 till 1990. End of Panchayat was made possible after the success of People's Movement led by Nepali Congress and United Left Front.

offered by the government to them.[16] Thus, there was no such visible dispute between the local people, government and the project construction authorities. However, with the restoration of democracy in 1990 and constitutional guarantee for freedom of speech, freedom of association and economic, social and cultural rights, people started to unite and resist against the violations of their rights. Likewise, the international instruments such as UN's Declaration on the Rights of Indigenous People and the ILOs Indigenous and Tribal Peoples Convention (popularly known as ILO 169) has given authority to the local people to raise their voice and demand for free, prior and informed consent before the confiscation of their land. As a result, local people demanded for information, public hearings for consultations on various issues such as land acquisition, compensation and individual and communal benefit sharing process. This has made the voice of locals stronger during the negotiation process. However, on some occasions, local people themselves are in disputes due to divergent interests and motivations in claiming their rights.

Although this chapter will later present three specific cases for the minute observation of disputes among hydropower stakeholders, this section briefly provides an overview of common hydropower disputes. In general, Nepal's hydroelectric conflicts are driven by vested political interests, trade-offs among hydropower stakeholders, power relations, societal norms, values and perceptions.[17] The disputes that arise during the construction of hydropower projects varies at local level and national level. At local level, the major source of disputes are on issues such as environmental impacts, land acquisition process, compensation for damages, compensation for public resources, benefit sharing, employment opportunities in the hydropower project construction, water sharing between the upstream-downstream communities and percentage of water flow in the river for drinking and irrigation purposes. In addition, the local stakeholders also raise their voice for proper dissemination of information and consultation on different issues that affects them. In contrast, national level stakeholders have often protested against the project on issues around broader cost and benefit of the project to the nation. National stakeholders are also

---

[16]Koirala, S. (2015). *Hydropower Induced Displacement in Nepal.* Dunedin, New Zealand: University of Otago.

[17]Upreti, B. R. (2007). Changing Political Context, New Power Relations and Hydro-Conflict in Nepal. In F. Rotberg & A. Swain (Eds.), *Natural Resources Security in South Asia: Nepal's Water* (pp. 15–45). Nacka, Sweden: Institute for Security and Development Policy.

concerned about the political aspects associated with the project such as whether the project violates any legal and constitutional provisions.

Due to dispute between stakeholders, projects are delayed and projects that are under construction halted for several days, weeks or even months. Lack of active consultation between local communities, project investors and the government regarding the construction of the project have worked as catalytic factor in creating dispute.[18]

MULTIPLE ACTORS AND MULTI-SCALE POLITICS IN HYDROPOWER CONSTRUCTION

The politics spinning around development of hydroelectric project is complex, as diverse stakeholders located at different scales attempt to influence the construction process as per their interest. Politics of scale has been referred as the nested hierarchy of territorial units with enclosed jurisdictional space which ranges from local, regional, national and global levels.[19] According to Dore and Lebel (2010) politics of scale refer to 'the tensions when and where actors cooperate, compete, or conflict as they endeavor to exercise their influence on the present and future of water resources use and further development'.[20] In many cases, the actors also jump scale and move from one scale to another to reframe the issues and influence the process as per their interests. In some cases, local activists jump scale and connect themselves with other national and international networks in order to bring the issues to the wider audience and pressure the project authorities to fulfil their demands.[21] Freeman (1984) defines stakeholder as group or individual who can affect, or is affected by development policies, programmes and activities. They are the range of interested parties who have different

---

[18]Voller, L. (2016, October 27). *Hydro Boom Sparks Violent Conflicts in Nepal.* https://danwatch.dk/en/hydro-boom-sparks-violent-conflicts-in-nepal/?cn-reloaded=1.

[19]Cox, K. R. (1998). Spaces of Dependence, Spaces of Engagement and the Politics of Scale, or Looking for Local Politics. *Political Geography, 17*(1), 1–23; Koirala, S. (2015). *Hydropower Induced Displacement in Nepal.* Dunedin, New Zealand: University of Otago.

[20]Dore, J., & Lebel, L. (2010). Deliberation and Scale in Mekong Region Water Governance. *Environmental Management, 46*(1), 60–80.

[21]Molle, Francois. (2007). Scales and Power in River Basin Management: The Chao Phraya River in Thailand. *The Geographical Journal, 173*(4), 358–373; Sneddon, C. (2002). Water Conflicts and River Basins: The Contradictions of Comanagement and Scale in Northeast Thailand. *Society &Natural Resources, 15*(8), 725–741.

MULTI-STAKEHOLDER HYDROPOWER DISPUTES ... 131

mindset, interests and power to influence the implementation process. This indicates the complexity of managing and resolving disputes of stakeholders situated in different scales. This section further illustrates major actors involved in hydro affairs in Nepal, their likely interest and agendas during the implementation of hydropower project.

Government: Hydropower construction is related to the utilization of the country's natural resources and production of electricity for the public good. Thus the government is directly or indirectly involved in formulating plans, policies and decision-making process on hydropower construction, depending on the scale of production.[22] The law of Nepal also grants liberty to the government to acquire, requisition or create encumbrances on its citizens' property for development purposes after paying compensation.[23] The government is equally liable to protect the rights of citizens and to safeguard its citizens the rights of its citizens who may be directly and indirectly affected by hydropower projects.[24] This indicates that the state holds power to acquire land, control entities through policies, laws, media campaigns, or in some instances even use threats and force as strategy for achieving its goal.[25] However, there are instances where governments compel to step back due to the pressure from locals and other local and transnational civil society actors.

Concerned Ministries and Departments of National and Federal Sates: In Nepal, hydropower construction requires approval and coordination of different ministries such as Ministry of Energy, Water Resources and Irrigation, Ministry of Land Management, Cooperatives and Poverty Alleviation, and Ministry of Forest and Environment. Different ministries and department have different objectives and are bestowed with different authorities. Hence, they may have different perspectives and interests in hydropower construction. So their approval, coordination and suggestions are necessary for effective implementation of the hydropower projects. Yet, at times

---

[22] Government of Nepal. (2001). *The Hydropower Development Policy, 2001.* Kathmandu: Ministry of Water Resources.

[23] Government of Nepal. (1990). *The Constitution of the Kingdom of Nepal 204/ (1990).* Kathmandu: Law Books Management Board.

[24] Government of Nepal. (2015). *The Constitution of Nepal.* Kathmandu: Law Books Management Board.

[25] Lebel, L., Garden, P., & Imamura, M. (2005). *The Politics of Scale, Position, and Place in the Governance of Water Resources in the Mekong Region.* Retrieved November 7, 2013 http://digitalcommons.usu.edu/unf_research/32/.

due to divergent interests and motivations, disputes arise between project developers and concerned ministries which affects the smooth and effective implementation of the hydroelectric project. For example, there was disagreement between 900 MW Upper Karnali Hydropower project developers and Department of Forest regarding the 273 hectares forest area the project would need for the project construction. Later, the dispute was solved after the government directed to provide the forest areas on lease to the project developers.[26]

Project Investors: Medium and large-scale hydropower construction is an expensive affair. Thus most of the governments and concerned authorities from developing countries take loans or grants from developed countries or multilateral and bilateral agencies to construct such large projects.[27] The investors provide loans to the developing countries but come with terms and conditions favouring their own interests such as profit-making, helping the development process and influencing as per their economic as well as geo-political agendas.[28] In Nepal, there are many cases where the locals, activists, civil society actors, political leaders have protested against the terms and conditions of project investors. For example, the Arun III hydropower project was opposed for its economic viability[29] and Pancheshwore Hydropower Project for social, environment, and geo-political aspect.[30]

Project Workers: These are the employees appointed by the hydroelectric construction company. Project workers play a vital role for the timely and effectively completion of the project. However, at times, interests of the project employee and employers do not match on issues around payment of wages, working hours, and other benefits entitled to receive as workers. In such situation, disputes arise, which also tends to disrupt the project

---

[26]The Kathmandu Post. (2017, October 19). *Upper Karnali Hydro Receives Forest Clearance*. Retrieved November 19, 2018 http://kathmandupost.ekantipur.com/news/2017-10-19/upper-karnali-hydro-receives-forest-clearance.html.

[27]Koirala, S. (2015). *Hydropower Induced Displacement in Nepal*. Dunedin, New Zealand: University of Otago.

[28]Ibid.

[29]Shrestha, R. S. (2009). Arun-III Project: Nepal's Electricity Crisis and Its Role in Current Load Shedding and Potential Role 10 Years Hence. *Hydro Nepal: Journal of Water, Energy and Environment, 4*, 30–35.

[30]Gyawali, D., & Dixit, A. (1999). Mahakali Impasse and Indo-Nepal Water Conflict. *Economic and Political Weekly, 34*(9), 553–564.

construction processes. For example, labours of Upper Tamakoshi project halted the construction works for days, demanding 500 project shares to each 800 project workers as a benefit of working for the project.[31]

Project Affectees: The construction of hydropower project or transmission line directly or indirectly affects the local communities residing in the project areas, its vicinity, en route to the transmission line. Physical displacement and loss of livelihood are two major impacts experienced by local communities, where physical displacement refers to the displacement of locals from their usual place of residents and loss of livelihood refers to being deprived from their usual livelihood practices during and after the construction of the project.[32,33] As these groups are affected by the project, the project attempts to compensate for their loss. However, in many cases the loss is quite massive and compensation provided by the developers may not satisfy the affectees. In such situation disputes arise. Further, many locals have ancestral ties with the place and are very much attached. Hence they refuse to leave the area, which also creates conflict between the project developers and the affectees. In Nepal's case, such disputes have occurred before or during the construction of hydroelectric projects like Kulekhani Hydropower Project, Kali Gandaki Hydropower Project and West Seti Hydropower Project.[34]

Project Beneficiaries: The project beneficiaries are people who benefit directly or indirectly from the construction of hydroelectric projects where direct beneficiaries are those who ultimately utilize the energy produced from the project and those who invest money on it. The indirect beneficiaries are those who are not considered as targeted beneficiaries of the project, however they benefit indirectly from the project construction. For example, people who come to the hydroelectric project area and start a new business or work in the project as a labour. Such beneficiaries of hydropower

---

[31] Harris, M. (2015). *NEA Forms Committee to Mediate Workers Strike at 456 MW Upper Tamakoshi Hydropower Plant.* Retrieved October 9, 2018 from Hydro World https://www.hydroworld.com/articles/2015/03/nea-forms-committee-to-mediate-worker-strike-at-456-mw-upper-tamakoshi-hydropower-plant.html.

[32] Penz, P., Drydyk, J., & Bose, S. P. (2011). *Displacement by Development: Ethics, Rights and Responsibilities.* New York: Cambridge University Press.

[33] World Commission on Dam. (2000). *Dams and Development.* The Report of the World Commission on Dams. London and Sterling: Earthscan.

[34] Koirala, S. (2015). *Hydropower Induced Displacement in Nepal.* Dunedin, New Zealand: University of Otago.

134 S. KOIRALA ET AL.

project may also go against the project implementation when they feel that the project does not fulfil their expected interests.

Civil Society: Civil society is a group of actors, networks and associations who come together with the objective of securing human rights, social justice, democracy and peace.[35] In the case of hydropower project, civil society often advocates for constructing only those projects which serves national as well as public interests. Civil society organizations pressurize the project developers and government to be accountable and transparent to its citizens.[36] The local, national and transnational civil society actors move from geographic scale and often advocate for the rights of the project affectees.[37] There are cases where civil society have been succeeded to influence the project developers and the government and there are cases where they have failed too.[38]

External Interest Groups: These groups are not directly related to the project, however, they have interests in the project activities. Media is one such interest group who are often concerned about the construction of hydroelectric projects or any other social and economic issues related to the project, theoretically with the intention of informing it to the wider audiences. The other external interest groups can be the neighbouring countries who are interested in the hydro affairs of the country. In the case of Nepal, India has shown its interest in Nepal's hydro affairs because river originating in and flowing from Nepal has been a huge source of water for India's irrigation projects. In addition, Nepal and India have inked a few agreements on sharing transboundary water resources which many Nepali citizens perceived to be in the favour of India.[39] Lately, Bangladesh has also shown its interest in Nepal's hydropower projects to meet their

---

[35] Dahal, D. R. (2001). *Civil Society in Nepal: Opening the Ground for Questions.* Kathmandu, Nepal: Center for Development & Governance.

[36] World Commission on Dam. (2000). *Dams and Development.* The Report of the World Commission on Dams. London and Sterling: Earthscan.

[37] Oliver-Smith, A. (2001). Displacement, Resistance and the Critique of Development: From the Grass Roots to the Global. In C. J. Wet (Ed.), *Development Induced Displacement: Problem, Policies and People.* New York: Berghahn Books.

[38] McDowell, C., & Morrell, G. (2013). *Displacement Beyond Conflict: Challenges for the 21st Century.* New York: Berghahn Books.

[39] Gyawali, D. (2001). *Water in Nepal.* Kathmandu: Himal Books.

huge electricity needs in the long run.[40] Likewise, China has also shown its interests in investing in Nepal's hydropower projects for profit-making and to have good geo-political relationship with Nepal.[41]

As we can understand from the above-mentioned interpretation that with the multiplicity of actors involved in the process of the construction of hydroelectric project, the dispute seems obvious for different reasons. In a democratic country like Nepal, it is very important to make different actors convince and address their key concerns before implementing a hydroelectric project. Yet, this process can be a hard hit struggle for all concerned actors, thus needs to face a number of disputes and address them with adequate measures.

HYDROPOWER PROJECTS AND DISPUTES

In this section, we present three hydroelectric projects (proposed and ongoing) that were trapped into different types of disputes before and during the commencement of the project.

Arun III Hydroelectric Project

The questioning of hydropower project in larger scale ascended with 201 MW Arun III Hydropower Project located in Sankhuwasabha district of eastern part of Nepal. Arun III was initially approved by the government of Nepali Congress in 1994 due to its viability to produce a large volume of electricity for meeting the local needs. The World Bank was the funding organization and the cost of the project was estimated to be around $764 Million.[42]

Initially, Arun III project was warmly welcomed by the then government, major political parties and local people. Later on, the project experienced unprecedented disputes and controversy both inside and outside of

---

[40]The Kathmandu Post. (2018). *Nepal, Bangladesh Agree to Build Hydro Projects.* Retrieved December 7, 2018 from *The Kathmandu Post* http://kathmandupost.ekantipur.com/news/2018-12-05/nepal-bangladesh-agree-to-build-hydro-projects.html.

[41]Koirala, S. (2015). *Hydropower Induced Displacement in Nepal.* Dunedin, New Zealand: University of Otago.

[42]Forbes, A. A. (1999). The Importance of Being Local: Villagers, NGOs, and the World Bank in the Arun Valley, Nepal. *Identities Global Studies in Culture and Power,* 6(2–3), 319–344.

Nepal through the networks of international and local non-governmental organizations (NGOs) as well as Arun III Project Concern Group. A lot of arguments were made and much was written about Arun III project, in particular on the economic viability of the project and adverse environmental impact that project could make among local communities. As an example to this, the then General-Secretary of Nepal Communist Party (UML) Madhav Kumar Nepal wrote a letter to the World Bank President on 18 October 1994 expressing their serious reservation regarding the cost–benefit and environmental aspect of the project.[43] At the same time, a hard hit anti-Arun III campaign was launched by international NGOs such as International Rivers Network, International Technology Development Group, Greenpeace International, Environment Defense Fund and Friends of the Earth in various financial capitals including Washington DC, Manila, Tokyo and Bonn.[44] In the meantime, International Institute for Human Rights, Environment and Development (INHURED International), a Kathmandu-based NGO also decided to monitor the environment and human rights aspect of Arun III project. Furthermore, INHURED International in partnership with Alliance for Energy and Arun Sarokar Samuha (Arun Concerned Group) decided to analyse the economic sustainability of the project. INHURED International also inspected the prerequisites of lending fund by the donors and its impact on the environment, social and the overall economy of Nepal.[45] Their report suggested that the construction of Arun III is not socially, economically and environmentally feasible for the country.

Based on the initial assessment, INHURED International requested Nepal Electricity Authority (NEA) to make the information and documents related to Arun III project public, but NEA rejected the request. In response to that, INHURED International along with other supporters of the campaign filed a Public Interest Litigation case at the Supreme Court of Nepal on 16 January 1994. The Supreme Court decided in favour of the petitioners.[46] Later on, INHURED International together with Arun

---

[43] Mahat, R. S. (2005). The Loss of Arun III. In *In Defence of Democracy: Dynamics and Fault Lines of Nepal's Political Economy* (p. x). New Delhi: Adroit.

[44] Ibid.

[45] Siwakoti, Gopal. (n.d.). *Nepalma Jalshrot tatha Jalbidhut bikash aviyan: Samikshyatmak Tipadhi*. Nepal.

[46] Clark, D., Fox, J. A., & Treakle, K. (2003). *Demanding Accountability: Civil Society Claims and the World Bank Inspection Panel*. Lanham, MD: Rowman & Littlefield.

Concerned Group and two local victims from the Arun Valley filed a case to the World Bank's inspection panel. After the inspection, the panel came to the conclusion that the project has violated the Bank's internal policies and does not comply with its own social and environment policy and guideline.[47] Later, the Bank's new President James Wolfenson consulted the panel and decided to withdraw from the project stating that the cost of the project is much higher than it's expected benefit. In order to resolve the controversies, Wolfenson also forwarded new power sector package for Nepal with smaller alternative approaches to meet Nepal's electricity needs.[48]

Concerned activists all around the world celebrated the cancellation of the Arun III Hydropower Project as a success of their activism. This victory also encouraged other activists to file case against the large-scale infrastructure projects inducing severe negative impacts on environment and local community. The case of Arun III showcases how civil society and hydro activists jumped scale to gain support from international community to pressure the government to cancel the Arun III project. Despite all these events, the story of Arun III has not ended yet.

The proponents of Arun III hydropower viewed that the cancellation of Arun III hydropower is one of the biggest loss of Nepal and one of the core reasons behind the electricity crisis that Nepal faced in the past two decades. Thus, the political leaders and bureaucrats who were in favour of the project were looking for an opportunity to retrieve Arun III. In 2008, the then Government of Nepal awarded to Satluj Jal Vidyut Nigam (SJVN) India Limited to the execution of Arun III project. SJVN produced a new Environment Impact Assessment (EIA) report and it was approved by the Ministry of Science, Technology and Environment, Nepal in 2015. The SJVN and the GON agreed to produce 900 MW of electricity of which Nepal will receive 21.9% free electricity from the project during the concession period. The remaining electricity produced from the plant will be exported to India. The project will operate on BOOT modality and will be transferred to Nepal after 30 years of power generation. The project will affect 269 families and physically displace 24 families. The affected locals opted for cash compensation over land compensation. The land acquisition

---

[47] Ibid.
[48] Ibid.

process started in 2016. Initially, the locals were not satisfied with the compensation price offered by the project developers. However, after several rounds of discussions between the affectees, compensation fixation committee and project developers, a new deal was offered and the disputes have been resolved. Tripartite meeting held between the locals, project developer and officials of Invest Board of Nepal (IBN), decided to give 65% more than the market price of the land which project acquired from the local people.[49] Additionally, the project has also assured to invest in the development of the local community through generating employment for the locals, boosting local industries, and fostering entrepreneurship. Project developer has also committed to provide robust community infrastructure and financial support through the construction of roads, bridges, schools, hospitals and community centres. Likewise, each household directly affected by the project will receive 30 units of free electricity per month. The project developer has allocated 1.6 billion rupees worth shares to the locals of the project area.

The case of Arun III hydropower showcases a unique story. The project, which was once highly disputed and controversial on environmental, economic, and social grounds was cancelled by the project investors, is now retrieved. Due to the larger capacity of the Arun III project than the past, the magnitude of social and environment impacts may remain higher. However, this time there hasn't been a strong voice against the construction of the project from local people as well as transnational NGOs. The economic viability of the project seems to be more reasonable than the previous plan of the government. The benefit that will be received by the locals is also better.

Since Arun III is a large-scale infrastructure project, the negative social and environment impact it will foster is unavoidable. Surprisingly, there is no dispute between project developers and civil society groups and local people on these two grounds. The political party, which was vocal about environmental and economic aspect of the project is now silent and welcomed the project wholeheartedly. This raises two concerns from the dispute resolution perspectives. First, the environmental and social concerns related to hydropower projects have become less significant over the period of time and over shadowed by the economic benefits of the project.

---

[49] My Republica. (2016). Smooth Sail for Arun III Land Acquisition Process. Retrieved November 25, 2018 from *My Republica* https://myrepublica.nagariknetwork.com/news/smooth-sail-for-arun-iii-land-acquisition-process/.

At present, not only the locals, but also the national level actors are content with the attractive compensation and benefit sharing packages that local people and country has received as per the agreement. This suggests that attractive economic benefits could be one of the strategies to minimize disputes that arise in the project construction process.

Second, the civil society networks working on hydropower issues have become weak in Nepal due to negative connotation attached with them as an anti-developmentalist and are also blamed for the electricity crisis in Nepal. Thus, civil society voice concerning environment and social impacts caused by the hydroelectric projects have become weaker. Likewise, the weak connection of national and transnational civil actors with local level activism has limited their ability to influence the local residents as per their interest. As a result, the locals of the project areas are less fragmented on issues around the construction of Arun III hydropower project. On the other hand, unlike the previous situation, political leaders have also shown solidarity towards the implementation of the project. This shows that the projects that have strong political support and less interference of civil society actors are less likely to have large-scale disputes. However, the long-term impact of such projects on environmental and social ground is still questionable. It can be argued that negative environmental and social impacts created by the project in the long run can be a great source of disputes between different stakeholders.

West Seti Hydropower Project

The proposed 750 MW West Seti Hydropower Project (WSHP) is a storage-based project located in Far-Western part of Nepal. This project was initially planned with the objective of exporting electricity to India.[50] The Government of Nepal and Snowy Mountains Engineering Corporation (SMEC), an Australian company, had signed a Memorandum of Understanding (MOU) in July 1994 for the construction of WSHP. The project components, including the dam site, reservoirs and the project site are located in four districts, namely, Bajhang,, Baitadi, Doti and Dedeldhura.[51] It was estimated that the proposed project activities conducted in

---

[50] Rai, K. (2005). *Dam Development: The Dynamics of Social Inequality in a Hydropower Project in Nepal.* Gottingen: Cuvillier Verlag.

[51] WSHL. (2007). *Summary Environmental Impact Assessment* (Environmental Assessment Report). Kathmandu: West Seti Hydro Ltd.

140 S. KOIRALA ET AL.

both downstream and upstream areas partly or fully affect the then 26 Village Development Committees (VDCs) of these four districts. In addition, the proposed reservoirs will submerge parts of 15 VDCs across the four project districts. It is estimated that the construction of project will displace 9968 people from 1190 households. The construction of transmission line associated with the WSHP would displace and affects additional people from the project areas, the estimated number yet to know.[52]

In 1997, the Nepal government granted license to SMEC for developing the WSHP under BOOT modality for 30 years. As per power purchase agreement between SMEC and West Seti Hydropower Limited (WSHL) and Power Trading Company India Limited (PTC), Nepal would receive revenue from the sale of power through energy and capacity royalties.[53] In addition, SMEC could either pay an amount equivalent to 10% of electricity or 10% of the total installed capacity of the energy that Nepal was supposed to receive free of cost.[54] The China Exim Bank granted the initial funding for the study of WSHP dam. Later, Asian Development Bank (ADB) approved the EIA and agreed to grant private-sector loans. China Machinery and Export and Import Corporation also joined the investing team. However, soon after the project was handed to SMEC, it was trapped into controversies and constantly opposed by different political parties as well as local, national and transnational civil society organizations for four major reasons.

The first reason for opposition was regarding government ignoring the provisions of Article 126 of the Constitution of Nepal 1990 while granting the license to SMEC.[55] Article 126 emphasized that any agreement or treaty on natural resources and their distribution must be ratified by two-third majority in the parliament before it can be enacted.[56] However, the then government did not bring this project into parliamentary discussion

---

[52] Koirala. S. (2015). *Hydropower Induced Displacement in Nepal.* Dunedin, New Zealand: University of Otago.

[53] Uprety, K. (2011). Nepal's West Seti Project Imbroglio: The Supreme Court Speaks. *Beijing Law Review*, 2(1), 17–24.

[54] Shrestha, H. M. (1997). *Pashim Seti Smriti Patra: Girdo Manastithile Rastrako Heet Gardaina.* Kathmandu: Deshantar Saptahik.

[55] Nepal already has its new constitution which is Constitution of Nepal 2015.

[56] Government of Nepal. (1990). *The Constitution of the Kingdom of Nepal 2047 (1990).* Kathmandu: Law Books Management Board.

MULTI-STAKEHOLDER HYDROPOWER DISPUTES ...   141

and approved it with two-third majority before signing the MOU with SMEC.

The second reason behind opposing the project was for ignoring and not charging India for downstream benefit. With the construction of WSHP, India would benefit by the increased flow of water in the West Seti reservoir which would help India to irrigate more than 1500,000 hectares of agricultural land.[57] The opponents argued that the government should adopt the principles set forth by the treaty between Lesotho and South Africa as well as Columbia River Treaty and charge India for the benefit augmented by the project.[58]

The third reason was related to unfair agreement on electricity sharing and electricity traffic rate provisioned in the agreement. Groups opposing the project argued that 10% electricity deal is not beneficial to Nepal, as the country is facing electricity crisis.[59] They further argued that Nepal has been purchasing electricity from PTC India Limited at around USD 9¢ per kilowatt per hour, whereas as per the agreement, Nepal was supposed to export electricity to India with an average tariff of USD 0.049¢ per kilowatt per hour, which is much lower than the price Nepal was paying to India.[60]

Another significant concern raised by the opponents and also supported by transnational civil society actors as well as local community was on the social and environmental aspects of the project. West Seti Concern Society (WSCS) formed by local people from the project area raised questions on several issues including compensation, resettlement, lack of transparency and inadequate information dissemination by the project developers. At one period of time, the opposition was so intense that the protests remained violent and the locals even set fire on the local office of the project located in Bajhang district. This made ADB susceptible to the successful implementation of the project. Further, SMEC was not able to conclude the power

---

[57]Shrestha, R. S. (2009). West Seti Hydroelectric Project: Assessment of Its Contribution to Nepal's Economic Development. *Hydro Nepal: Journal of Water, Energy and Environment,* 5, 8–17.

[58]Ibid.

[59]Shrestha, H. M. (1997). *Pashim Seti Smriti Patra: Girdo Manastithile Rastrako Heet Gardaina.* Kathmandu: Deshantar Saptahik.

[60]Shrestha, R. S. (2009). West Seti Hydroelectric Project: Assessment of Its Contribution to Nepal's Economic Development. *Hydro Nepal: Journal of Water, Energy and Environment,* 5, 8–17.

purchasing deal with India. As a result, in 2010, ADB and China Machinery and Export and Import Corporation withdrew from the project.[61]

Later in July 2011, the government scrapped the license of the project which SMEC had obtained and granted the project to China Three Gorges Company (CTGC) under the new terms and conditions. As per the MOU, CTGC and NEA will finance 75 and 25% of the shares respectively and the local communities are entitled to obtain 2–5% of the shares from CTGC owned shares. As per the new provision, the project is planned to build for national consumption. The new agreement is appreciated by most of the concerned actors and welcomed by locals including the civil society actors who opposed the project previously.[62] The project was supposed to commence the construction work in 2014 and was expected to complete it by 2019. However, the work has not begun yet and has been trapped into technical controversies between the Chinese investors and the IBN on issues around land acquisition, resettlement, rehabilitation of the local people and arranging power transmission line from remote project area to the Kathmandu.[63]

The case of WSHP demonstrates wide range of disputes between key stakeholders of the project. At national level, the opponents of the project were political party leaders, civil society and hydro activists living in Kathmandu. The major issue raised by these actors were regarding unfair agreement between the government and project investor on power trading agreement with India, downstream upstream benefit sharing between Nepal and India, violation of the constitutional provisions while signing the MOU with project investors, and paying less attention towards safeguarding social and environmental agenda while commencing the project. Another layer of dispute is observed between the government and SMEC regarding the project commencement modality, project financing, and power trading agreement with India. At the local level, the opponents of the project were local civil society actors, local political leaders and local people themselves. They opposed the project developers for not providing

[61] Rai, D. (2012). *In Search of Light*. Retrieved October 19, 2018 from http://archive.nepalitimes.com/blogs/thebrief/2012/03/08/in-search-of-light/.

[62] Koirala. S. (2015). *Hydropower Induced Displacement in Nepal*. Dunedin, New Zealand: University of Otago.

[63] Ghimire Y. (2018). China Eyes Exit, Nepal's West Seti Hydropower Project in Jeopardy. *South China Morning Post*. https://www.scmp.com/week-asia/geopolitics/article/2161968/nepals-west-seti-hydropower-project-jeopardy-china-eyes-exit.

adequate information regarding project, social and environment impacts that project will foster, and compensation, resettlement and rehabilitation package for the affected people. With these disputes at local and national level, gestation period of the project has become longer and eventually government had to cancel the project license of SMEC. This illustrates that, dispute between key stakeholders of the project can delay in project implementation. It can also be a major factor behind the termination of the project. In this case, the government cancelled the license of project developers which served as a tool to resolve the ongoing dispute on WSHP, yet this cannot be considered as durable solution to hydropower dispute.

Although the government has granted WSHP license to a new company CTGC, the dispute between them on issues around land acquisition, resettlement, rehabilitation of the local people and arranging power transmission line is dragging the project commencement date behind. Government and project investors are yet to come up with strategies for addressing these issues. However, the implementation of WSHP still seems to be challenging given the number of locals displaced and affected by the project. Local people's high-level of expectations with the implementing agencies on the matter of compensation and benefit sharing can be a great source of dispute for a long period of time. This suggests that a real dispute particularly on issues around compensation, rehabilitation and benefit sharing is yet to come. This dispute is going to be even complex as the number of affectees and displaces of WSHP is relatively larger, thus dealing with a larger number of people can be a daunting challenge. Likewise, people from the WSHP area have waited long to realize this project into reality and their expectations from this project may be higher in terms of compensation, rehabilitation and benefit sharing, thus it can also be a great source of dispute in the days to come.

Upper Tamakoshi Hydropower Project

The 456 MW Upper Tamakoshi Hydropower Project is a run-of-the-river hydroelectric project located in Lamabagar of Dolakha district of Nepal. It is built in the Tamakoshi River, nearby Nepal–Tibet border. In 2007, NEA established Upper Tamakoshi Hydropower Limited (UTKHPL), an autonomous company for the execution of this project. The project is contracted to a Chinese company, Sinohydro Corporation Limited for civil construction works. The project has been financed by domestic financial institutions and companies. As per the agreement among the investors,

NEA, Nepal Telecom, Citizen Investment Trust and Rastriya Beema Sansthan owns 41, 6, 2 and 2% of the company's shares, respectively. Likewise, the general public and residents of Dolakha own 15 and 10% of the shares, respectively. The contributors of Employees Provident Fund, staffs of NEA, UTKHPL and other financial institutions providing loans have distributed 24% of the project share.[64] The project construction work started from 2011 and was expected to be completed by 2018.[65] The electricity produced under this project will be consumed within the country.

One unique feature of the project is the involvement of the Upper Tamakoshi Peoples Concern Committee (UTPCC) even prior to the Upper Tamakoshi's operation started. Formed in 2001, UTPCC comprised of then Members of Parliament from Dolakha district, representatives of political parties, local civil society representatives and distinguished personalities from the district. The UTPCC was formed with the motive of supporting the implementation process of Upper Tamakoshi Project through the level of public residing in Dolakha district. Out of many project supporting activities organized by UTPCC, a highly remarkable one was a promotional workshop in Kathmandu on 2 February 2002 with the participation of concerned governmental and non-governmental representatives. The workshop was participated by the then Finance Minister, Water Resources Minister, ex-Ministers, representatives of major political parties, Nepali business community, civil society members and media persons. The participants of the workshop appreciated the efforts made by local people to build consensus among relevant stakeholders regarding the need for the project and minimize the level of disputes before commencing the project. Later on, through the collective efforts of UTPCC, NEA, National Planning Commission and local political leaders, Norwegian Government was convinced to fund the feasibility studies including EIA of the project.[66]

The EIA illustrated that the project will incur negligible physical and environmental impact, but a moderate level of biological, social, economic and cultural impacts among local communities living in the area.

---

[64] UTKHL. (2011). *Upper Tamakoshi Hydroelectric Project Introduction and Present Status*. Upper Tamakoshi Hydroelectric Project Limited, Kathmandu, Nepal.

[65] The Kathmandu Post. (2016). *Upper Tamakoshi Project: Completion Deadline Pushed Back Again*. Retrieved December 1, 2016 http://kathmandupost.ekantipur.com/news/2016-12-22/completion-deadline-pushed-back-again.html.

[66] Neupane, S. (2016). *Project, People and Consent Managing Local Expectation in Hydropower Development*. Kathmandu: Center Department of Public Administration.

The project acquired 182 hectare of land of which 66 hectare is agriculture land, 78 hectare forest land, and 38 hectare barren and cliff land. A total of 276 households were directly affected by land acquisition for construction of project related infrastructure including construction of access road track and improving the existing road.[67] Among the affected households, 21 households have been classified as being Seriously Project Affected Families, of which only 14 families were relocated while others lost their land and property. One school located in project sites was also relocated in safer area. The project affectees were given cash compensation for their loss. There were some issues raised by the locals regarding compensation, and other social impacts, however, their disagreements did not take a conflicting shape, as their concerns were more or less solved in a right away. In addition, there were disputes on access road alignment in Singati Bazaar. The locals wanted the road to be constructed through the main market area, while the project feasibility study considered the road to bypass the market. The UTPCC mediated the dispute between the locals and project developers and the problem was solved.[68]

A major dispute that halted the ongoing construction of UTKHP was around local share distribution to the community and project workers. The project construction work was halted several times by the affectees and the project workers. The project affectees through their protests had mainly demanded (a) additional units of shares to the locals of Dolakha district, (b) re-categorization of project affected areas as highly affected, very affected, affected and less affected, and (c) locals in the affected area to be given stake in general public share of 15% as well.[69] The UTPCC mediated the dispute between the project developers and the agitating locals. Later changes were made to the distribution pattern of 10% share that was allocated to the locals of the project area. The project area was categorized as severely affected area, affected area, and rest of the district

---

[67] MoEST. (2005). *EIA Report of Upper Tamakoshi Hydroelectric Project.* Kathmandu: Department of Electricity Development.

[68] Neupane, S. (2016). *Project, People and Consent Managing Local Expectation in Hydropower Development.* Kathmandu: Center Department of Public Administration.

[69] Neupane, S. (2016). *Project, People and Consent Managing Local Expectation in Hydropower Development.* Kathmandu: Center Department of Public Administration.

and the numbers of shares were distributed to the locals depending on the areas they reside.[70]

The UTPCC was also involved in mediating the protest of project workers. Around 850 Nepali project workers[71] protested against the project developers and halted the construction work with their demanded of allocating 500 unit of shares to each worker. As there was no such provision mentioned in any official documents that the workers contracted by Sinohydro Cooperation Limited will receive shares of the project. The shares were meant to be allocated only for the staffs of the companies investing in the project. Several rounds of discussions and consultations were held between the agitating workers and project developers with the support of mediation team, yet they could not come to any conclusions. The process of halting the construction work and re-starting it again after the assurance of fulfilling their demands went on for some months.[72] Later on, two mega earthquakes that hit Nepal in April and May 2015 diverted everyone's attention from the ongoing dispute between the project workers and developers. The agitating workers also dispersed to handle their personal problems after the earthquake. The project also had some damage during an earthquake.[73] In fact, the earthquake played an important role to weaken workers' voice.[74]

Upper Tamakoshi is a unique case to observe dispute resolution practices. From the above-mentioned description it can be said that no major disputes observed among the key stakeholders before the commencement of the project. One major feature of this project is that the local people and local stakeholders united under the concern committee played an important role to minimize the level of dispute and effectively mediate the existing disputes with minor chances of reoccurrence. In many cases, local

---

[70] IFC. (2018). *Local Shares: An In-Depth Examination of the Opportunities and Risks For Local Communities Seeking to Invest In Nepal's Hydropower Projects*. Washington, DC: International Finance Corporation.

[71] Sinohydro Cooperation had appointed both Chinese and Nepali workers for the construction of the project.

[72] The Kathmandu Post. (2016). *Upper Tamakoshi Project: Completion Deadline Pushed Back Again*. Retrieved on December 1, 2016 from http://kathmandupost.ekantipur.com/news/2016-12-22/completion-deadline-pushed-back-again.html.

[73] The project suffered the loss of 10 m per day due to the obstruction of work by the project workers.

[74] Neupane, S. (2016). *Project, People and Consent Managing Local Expectation in Hydropower Development*. Kathmandu: Center Department of Public Administration.

actors are the one who often have conflicting relationships with the government and project investors. Whereas in the case of Upper Tamakoshi project, locals are the ones bringing all parties together and building consensus for the smooth completion of the project. Being a locally funded project built with the motive of domestic consumption of electricity, Upper Tamakoshi project had fewer opponents. The project did not have to go through the criticism of compromising the national interests and fostering more benefits to the project investors. Since the project did not have huge implications on environment and social grounds, nationals and transnational civil society actors also took the project positively. However, if the project would have been granted to international investors solely engaged with profit-making motives and had imposed their terms and conditions, there would be less chances for smooth implementation of the project and more chances of opposition from local, national, and transnational actors. Further, low level of dispute in the case of Upper Tamakoshi project is concerned with relatively fair distribution of shares among investors and local communities. In addition, the share distribution process was transparent from the very initial phase of the project. Likewise, the project had relatively less number of affected families and had provided cash compensation as per their expectations. The mediating role of concern committee either during small disputes or big had also played an important role to reduce the level of hostility. Moreover, natural disaster, sometimes plays an important role to make the dispute vanish from the scene. While in the case of Upper Tamakoshi project, the dispute between the workers and project developers vanished due to a mega earthquake that badly affected Dolakha district as well as the project.

Discussion and Analysis

The above-mentioned descriptions of three hydroelectric projects provide some important insights into hydropower disputes and their resolutions in Nepal. One important insight is related to differences in disputing issues between local stakeholders and national stakeholders. The local actors are found in disputes with the government and project investors on issues around potential individual benefits such as compensation and long-term financial gain they may obtain from the construction of a hydroelectric project. Disputing issues raised by the locals in all three above-mentioned hydroelectric projects are related to specific problems likely to be faced by them during and after the construction of hydropower project.

That is why, local stakeholders have always negotiated with project developers for better compensation packages, employment opportunities, community development initiatives and benefit sharing from the project.

In contrast, national and transnational civil society actors have opposed the government and project investors on broader issues such as transparency and fairness of project contracting process, environmental protection, inadequate upstream-downstream benefit sharing, and the violation of constitutional provisions as well as international laws and policies for protecting local communities' socio-economic and cultural rights. The other issues raised by civil society actors in Nepal is related to its water relationships with India. If large-scale hydropower projects have any connection with India then national civil society actors always bolster their voice for fairness in benefit sharing as well as make project mutually beneficial to both countries.

A number of reasons can be given for differences in disputing issues of local and national non-state actors with their relationships with the government and project investors. Inadequate flow of information at the local level regarding the ongoing debate and disputes at the national level often restricts local actors to raise their own specific issues. For example, many local people residing in WSHP areas were unaware of national debate on issues such as who is going to produce the electricity, who will benefit from the electricity production, for whom they are paying the cost of displacement, and other long-term consequences of the project.[75] As a result, their resistance was more concerned about their rights and the financial security of their future.

This study also demonstrates that disputing issues related to hydropower projects have changed over time. For example, when Arun III project was first introduced in the 1990s, the core issues of dispute were environmental protection and economic viability of the project, whereas these issues were no more in the limelight when the same project was reintroduced in 2014. Likewise, when the WSHP was initially granted to SMEC, many disputing issues were related with ignoring national benefit and adverse impact the project will make on project-affected communities. However, with the changes in the modality of the project, the WSHP was welcomed by many local and national stakeholders who opposed the project in the beginning. Although the project will displace a huge number of locals and will still have

---

[75] Koirala, S. (2015). *Hydropower Induced Displacement in Nepal.* Dunedin, New Zealand, University of Otago.

negative environment and social impact, these issues have got less attention than they used to be a decade ago. One reason behind changes in disputing issues is influenced by the global wave of social and environmental movement. In the 1990s, social and environmental movements against large-scale infrastructure projects that trigger negative impacts were so powerful globally and had active networks in many countries that brought together the voice of activists working for the rights of local communities. However, such movements at present have been weak around the globe and in Nepal as well. Another important dimension in this regard can be considered as the shifting needs of local communities and nation as well. The growing electricity crisis and demand throughout the country has compelled people to accept the large-scale hydropower project even if it might have some negative impacts in future. As a result, the civil society actors and activists who advocate against such projects are less heard and supported by the people as well. Likewise, local people are less concerned about issues that do not directly and immediately impact them, whereas more concerned about issues that has direct impact on their livelihood and economic sustainability once they are forced to displace from their ancestral land.

Another important dimension this study suggests that hydroelectric projects that are nationally funded are less disputed than externally funded. In the context of Nepal, people and the local civil society actors evaluate externally funded projects from the nationalistic point of view, thus there are lots of suspicion among them whether or not it benefits the people and the nation. Hydropower projects particularly become more sensitive when the project is executed by Indian companies and electricity is planned to sell in India. Whereas in the case of nationally funded project, there is no such debates and discussions observed in terms of ultimate gain the project provides. The reason WSHP and Arun III have been delayed for many years in their commencement and became highly contested is because of external involvements in the construction of the project, whereas Upper Tamakoshi Hydropower Project came into execution relatively faster because of its entire funding through national investors and local people.

In all three cases, compensation and benefit sharing among the local and project-affected communities have been a source of dispute as well as an effective tool to resolve those disputes. Local people who resisted against the project had demanded adequate compensation for the confiscated land, damage of property and cost of displacement from their usual place of residence. Likewise, local communities also raised their voice for individual and collective benefit sharing in return to their sacrifice for the execution of the project. Upper Tamakoshi Hydropower Project has been so successful in resolving compensation and benefit sharing issues through constant dialogue, consultation and negotiation among key stakeholders. Such disputes have been resolved even in the case of Arun III when this project was reconvened in 2014. These issues are yet to be resolved in the case of WSHP.

This study also suggests that dispute often becomes a major factor for delaying the commencement of hydropower project, but at the same time it also gives opportunities to crack new and better deals. Arun III can be a unique case in this regard where delay in implementing this project has been economically beneficial to local people as well as the nation, excluding the environmental and social impact the project will foster. The case of WSHP is also similar. The ongoing disputes around this project encouraged the government to change the construction company and come up with a better deal that favours the nation. Yet, Nepal has still a long way to go to resolve compensation and benefit sharing issues in relation to the West Seti Hydropower Project.

CONCLUSIONS AND RECOMMENDATIONS

The politics revolving around water resources is complex, whether it is related to the construction of hydropower projects, irrigation channels or sharing water with neighbouring countries. Diverse stakeholders and interest groups residing at different scales attempt to influence the water projects as per their own interests from the very inception of any project, which eventually becomes a major factor to stimulate disputes between project stakeholders. Hydropower projects in particular are found more prone to dispute due to the multiplicity of actors involved in the project construction process. By nature, large-scale hydropower projects have potential to displace large number of people from the project areas. It can also be a contributing factor to the loss of physical property as well as usual livelihood opportunities of local people. Such projects being constructed in ancestral land, people desire to have some sustainable financial gain at the cost of their displacement and being impacted socially as well as environmentally. Hence, they try their best for better benefit package whether by advocating or bargaining or protesting against the project. If their demands are not met easily then they become confrontational and halt the project construction process. This result delays in the construction of the project and increases the cost of the project.

While talking from the government and project investors' perspectives, they desire to maximize the profit and get things done with nominal hurdles during the project construction process. Thus, their motives are to give less to the local people and workers involved in the project. In such a scenario, dispute between these actors seem obvious. The main question is how to resolve dispute constructively, so that the large-scale hydropower projects can be completed on time; people affected by the project are satisfied with what they have received in return to their sacrifice; there would be less damage of environment; social fabric between communities will remain intact; and economic gain of the project will be shared equitably among the local community, government, and project investors. However, in reality, the situation cannot be the same as expected. With the presentation of three cases mentioned above, our conclusion is that, dispute in relation to large-scale hydropower projects is a bitter reality. Dynamics of hydropower dispute has been changing from past to present. In the past, disputes were heavily focused on environmental, social as well as economic aspects of the project. Nationalistic issue was also a dominant discourse in the past. However, at present, disputes are less concentrated on environmental issues. Whereas economic issues such as compensation to the affected communities and benefit sharing among local communities as well as project construction workers have been a dominant discourse and core to resolve and reduce all sorts of disputes in relation to hydropower projects. In contrast, the government has no concrete policies or guiding document for giving compensation[76] and sharing benefits among the local communities and affected population. Whatever is happening at present is based on ad hoc basis and largely depends on the negotiation capacity of disputing parties.

This study also clearly suggests that taking local actors into confidence is very important for the constructive resolution of hydropower disputes. Local actors' initiations in resolving disputes in the case of Upper Tamakoshi Hydropower Project can be taken as a good example in this regard. The benefit sharing mechanism adopted by the project was exemplary in gaining public support for the project. Public's stake in the project made them feel a sense of ownership and responsibility towards the successful implementation of the project.

---

[76]Land Acquisition Act 1977 has some provisions for giving compensation to the confiscated land. Other than that no compensation policy is in place.

Likewise, along with EIA and Social Impact Assessment (SIA), comprehensive conflict analysis of each hydropower project is key to make the concerned actors come with adequate dispute resolution mechanism and strategies beforehand for addressing possible disputes that may arise before, during, and after the commencement of the project. Additionally, constructive interaction among major stakeholders is essential to understand the concerns and interests of each stakeholders and find a common ground for all. Such meaningful interaction can ultimately create an enabling environment for timely completion of hydropower project without significant disputes.

This study also suggests the need for nationally owned benefit sharing guidelines or policy for resolving disputes associated with small, medium, and large-scale hydropower projects. This policy should be designed through a wider consultative process among hydro stakeholders and should speak clearly on what the locals can demand from the developers, what the developers are supposed to give back to the locals, and what role the government is supposed to play in terms of mitigating conflicts of interests between the two former groups, should they arise.[77] If comprehensive benefit sharing policy is not possible then, each project before their commencement should come up with benefit sharing plan through a wider consultation among the key stakeholders of the project and adhere to such policy while executing the project.

Last but not the least, EIA, SIA must be in the topmost priority of the government for sustainable solutions to forthcoming environmental and social disputes in relation to large-scale hydropower projects.

---

[77]Shrestha, A. (2017). Impediments to Hydropower Development in Nepal. Retrieved from *The Himalayan Times* on January 8, 2017 from http://samriddhi.org/news-and-updates/impediments-to-hydropower-development-in-nepal/.

Is Pakistan Running Dry?

Zofeen T. Ebrahim

In an article[1] on water shortage in Pakistan, Dr Arif Anwar, principal researcher with International Water Management Institute's (IWMI's) Pakistan office had predicted increased 'anarchy' in Karachi and a 'mafia more insidious than the tanker owners'[2] who will exploit the situation. 'I would love to paint a rosy picture of Karachi with beautiful waterways, a clean beach, and fountains', he said, 'but we need to change course drastically for that to happen'.[3]

---

[1] Ebrahim, Z. T. Is Pakistan Running Out of Fresh Water? *MIT Technology Review.* http://www.technologyreview.pk/is-pakistan-running-out-of-fresh-water/.
[2] Ibid.
[3] Ibid.

Z. T. Ebrahim (✉)
Karachi, Pakistan

© The Author(s) 2020
A. Ranjan (ed.), *Water Issues in Himalayan South Asia*,
https://doi.org/10.1007/978-981-32-9614-5_7

The 2018 National Water Policy of Pakistan states that from a 5260 m^3 surface water availability back in 1951, it has been reduced to 1000 m^3 in 2016. If things remain the way they are, it is likely to drop to 860 m^3 by 2050.[4]

But long before the scary proclamation, Pakistan Council of Research in Water Resources (PCRWR), an agency under the Government of Pakistan, said that the country had crossed the 'water stress line' in 1990 and in 2005 it crossed the 'water scarcity line'.[5]

The national policy on water therefore gives a clarion call for 'rapid development and management of the country's water resources on a water footing'.[6] In comparison to water, power has been on the top rung on the national agenda. Power shortage pull crowd out on the street. In fact, Pakistan Muslim League (Nawaz) came to power (2013–2018) on the promise that it would put an end to power outages if people voted for it and analysts had said back in 2016 that if it could eradicate 'load shedding' its chances of securing another term in office would increase significantly.[7]

'Water is a significantly poorer cousin to power. Its effects are slower, but far more lethal. If the wells run dry, food will disappear from the table!' warns Anwar, in an interview to the writer.[8]

But things are not that simple and a reductionist approach, say experts, will not address Pakistan's water management problems. For example, they say, a dam alone won't solve all of its water woes, nor will water pricing alone, nor will rehabilitating more canals or institutional reforms.

---

[4] Government of Pakistan, Ministry of Water Resources. *National Water Policy 2018.* http://www.ffc.gov.pk/download/AFR/National%20Water%20Policy%20-April%202018%20FINAL.pdf or https://mowr.gov.pk/wp-content/uploads/2018/06/National-Water-policy-2018-2.pdf. Accessed August 12, 2018.

[5] Buhne, N. (2017, November 15). Water Insecurity. *Dawn.* https://www.dawn.com/news/1370550. Accessed March 5, 2019.

[6] Government of Pakistan, Ministry of Water Resources. *National Water Policy 2018.* http://www.ffc.gov.pk/download/AFR/National%20Water%20Policy%20-April%202018%20FINAL.pdf or https://mowr.gov.pk/wp-content/uploads/2018/06/National-Water-policy-2018-2.pdf. Accessed August 12, 2018.

[7] Pakistan on Track to End Power Shortages Within Two Years—ADB. (2016, June 21). *Reuters.* https://www.reuters.com/article/pakistan-energy/pakistan-on-track-to-end-power-shortages-within-two-years-adb-idUSL8N19C3M4. Accessed July 15, 2018.

[8] Author interviewed Dr Arif Anwar over an email.

CLIMATE CHANGE

In addition, the water scarcity issue is compounded by a runaway population and the uncertainty posed by climate change that Pakistan is witnessing. The unpredictable precipitation is causing flash floods in the north and prolonged drought in the South.

The Global Climate Risk Index 2019, which released its report at the Climate Summit (COP24), at the Polish city of Katowice, in December 2018, placed Pakistan at number eight among the top ten countries that have been bearing the brunt of extreme weather since the past two decades—from 1998 to 2017.[9] The country has suffered a loss worth $384 million last year due to extreme weather events.

The warming weather is leading to formation of more glacial lakes. According to the United Nations Development Programme (UNDP), currently there about 3044 glacial lakes in the Gilgit-Baltistan (GB) and Khyber Pakthankhuwa (KP). Of these, about 33 lakes are the result of sudden and hazardous glacial lake outburst flooding (GLOF).[10] A GLOF can release around millions of cubic metres of water and debris which could lead to loss of lives, property and livelihoods among the remote and impoverished mountain communities. It is estimated that over 7.1 million people in GB and KP are vulnerable to such phenomena.[11]

Notwithstanding the grim statistics on water scarcity and warnings by the country's top researchers for reversing environmental degradation and ensuring equitable allocation of water, the signs of water shortages can clearly be seen even by lay people—in the form of water disputes, polluted and contaminated water supply and depletion.

---

[9] Eckstein, D., Hutfils, M.-L., & Winges, M. (2018). *Global Climate Risk Index 2019: Who Suffers Most from Extreme Weather Events? Weather-related Loss Events in 2017 and 1998 to 2017.* https://www.germanwatch.org/sites/germanwatch.org/files/Global%20Climate%20Risk%20Index%202019_2.pdf. Accessed January 5, 2019.

[10] *Scaling-Up of Glacial Lake Outburst Flood (GLOF) Risk Reduction in Northern Pakistan.* United Nations Development Programme, Pakistan. http://www.pk.undp.org/content/pakistan/en/home/operations/projects/environment_and_energy/project_sample.html.

[11] *Scaling-Up of Glacial Lake Outburst Flood (GLOF) Risk Reduction in Northern Pakistan.* United Nations Development Programme, Pakistan. http://www.pk.undp.org/content/pakistan/en/home/operations/projects/environment_and_energy/project_sample.html.

RUNAWAY POPULATION

Looking at the mighty Indus now which has become a mere dribble and the parched delta where once lush rice paddies ruled, one would think water has indeed become scarce. But that is far from true say experts. Many water professionals use the Falkenmark Water Stress Indicator—one of the most common scale to calculate the volume of per capita availability of freshwater. It determines the level of water scarcity. Anwar does it too. He says the decrease in water availability in Pakistan is, mainly, due to increase in population, and not decrease in availability of waters. 'Pakistan is heading below the water-scarce threshold because of population increase not necessarily because the volume of water in the country has decreased', he said.[12]

The same is corroborated by Pakistani environmentalist, and former diplomat, Shafqat Kakakhel, who has served as the deputy executive director of the UN Environment Programme (UNEP) and as UN Assistant Secretary General. He says Pakistan's 'runaway population growth is the single biggest reason for the drastic reduction in our per capita availability of water'.[13]

According to a 2015 IMF report the demand for water will reach 274 million acre feet (MAF) by 2025, but supply will remain stagnant at 191 MAF, resulting in a demand–supply gap of approximately 83 MAF.[14]

The 2016 census put the country's population at 207.7 million, and is predicted to cross the 395 millionth mark on the Pakistan's 100th year of independence in 2047, if the population growth is not controlled, and it remains as business as usual.[15] And as with the exponential rise in population, so will the demand for water. A 2015 IMF report states the demand for water in Pakistan is projected to reach 274 MAF by 2025, while supply

---

[12] Author interviewed Dr Arif Anwar over an email.

[13] Author interviewed Shafqat Kakakhel over an email.

[14] International Monetary Fund. (2015). *Issues in Managing Water Challenges and Policy Instruments: Regional Perspectives and Case Studies.* https://www.imf.org/external/pubs/ft/sdn/2015/sdn1511tn.pdf. Accessed June 29, 2018.

[15] International Monetary Fund. (2015). *Issues in Managing Water Challenges and Policy Instruments: Regional Perspectives and Case Studies.* https://www.imf.org/external/pubs/ft/sdn/2015/sdn1511tn.pdf. Accessed June 29, 2018.

is expected to remain stagnant at 191 MAF. This will result in an increase in demand–supply gap of approximately 83 MAF.[16]

Still there are climate deniers who refuse to believe there is a serious water scarcity issue. With three snow-capped mountain ranges—the Himalayas, the Hindu Kush and the Karakoram—that Pakistan is ensconced in and which together span 11,780 km^2, accounting for 7259 glaciers (containing 2066 km^3 of ice); they believe the country is blessed with an infinite supply of water. These glaciers feed the Indus river and its about 1.12 million kilometres wide basin area. Of the total basin 47% is in Pakistan, 39, 8, and 6% are in India, China, and Afghanistan, respectively.

To-Do List

The first ever National water Policy that was approved in 2018 is to guide each of the four provinces in developing and formulating their own plans for water conservation, water development and water management. Pakistan has a federal government but after the passage of the 18th Constitutional Amendment, the four provinces—Punjab, Sindh, Khyber Pakhtunkhwa and Balochistan—enjoy a considerable degree of autonomy.[17]

The policy is based on the concept of integrated water resources management to promote sustainable consumption, judicious and equitable utilization of available water.

It covers more or less all water-related issues including improving the quality of the available freshwaters to meet the domestic, agricultural, energy, security and environmental needs; improve the urban water management system by increasing its efficiency; raise public awareness towards wastage of water and, at the same time, promote conservation practices; develop and increase the share of hydropower in the energy mix.

The policy makes reference to treatment and reuse of wastewater—for domestic, industrial and agriculture; upgrading water sector information systems; improving watershed management by conservation of soil, catchment area treatment; preservation of forests and increasing forest cover

---

[16] International Monetary Fund. (2015). *Issues in Managing Water Challenges and Policy Instruments: Regional Perspectives and Case Studies.* https://www.imf.org/external/pubs/ft/sdn/2015/sdn1511tn.pdf. Accessed June 29, 2018.

[17] Eighteenth Amendment to Constitution of Pakistan. (2010). www.comparativeconstitutionsproject.org/files/pakistan_2010.pdf. Accessed January 12, 2002.

and focusing on flood and drought management to minimise damages; promote appropriate technologies for rain water harvesting in both rural and urban areas; regulating abstraction of groundwater so that aquifers can be recharged; devising a strategy for adequate pricing for water used for irrigation; operation and maintenance of irrigation system, promote research on water resources related issues; preserving the delta by providing regular and sufficient supply of water and implementation of the 1991 Water Apportionment Accord in letter and spirit.

DOCUMENTS, STUDIES, REPORTS ON WATER GATHERING DUST

To address the water issues, over the years, several researches and studies have tried to find solutions to water distribution, sharing and conservation. But nothing has worked, it seems.

Water Apportionment Accord

Among those is a piece of legislation called the Water Apportionment Accord 1991, for allocation of the 117.35 million acre feet (MAF) of water in the Indus Basin among the provinces (although there may never be that exact amount and the accord does not say what happens if the volume is less).[18]

The accord may be intact, but there remains heartburn between provinces, specially between Sindh (lower riparian) and Punjab (upper riparian) on water management in the Indus basin. To give an example, Daanish Mustafa, a Reader in Human Geography at King's College, London, says, the Chashma Jehlum Link and Trimmu-Panjnad Link canals are to be operated only with the consent of the chief minister of Sindh, but they are, in fact, quite arbitrarily operated by—what the Sindh perceive to be, and with some truth—a Punjab-dominated Indus River System Authority (IRSA), without the required consent of the Sindh's chief minister.

Furthermore, the old mindset continues to plague water experts in Punjab who see every drop of water that reaches the Indus delta and into the Arabian Sea a colossal loss.

---

[18] Government of Pakistan, Ministry of Water & Power. (1991). *Water Apportionment Act.* https://mowr.gov.pk/wp-content/uploads/2018/05/Water-Accord-1991.pdf. Accessed June 25, 2008.

Mustafa, who has written a book on water resources management in times of climate change, says Sindh does not trust the system. Yet there is a system whereby they can go to the Council of Common Interest (CCI) with a complaint but have never done that. Sindh's complaint is more about the operational aspects of water policy. Even the Sindhi nationalists supported the water accord as the basis of water distribution among the provinces. However the issue is about day to day operations which the CCI cannot look after, and here lies the devil![19]

But how has it helped provinces? According to Mustafa, it has been useful in the sense that it has provided a framework for interprovincial water distribution and a mechanism through which interprovincial water grievances can be articulated. The IRSA was established to regulate and monitor the distribution of the waters from Indus in accordance to the formula accepted in the 1991 accord among the provinces and to provide matters related therewith and ancillary thereto is a good forum to negotiate any water disputes, not that they ever do. They seem to just vote and the majority carries the day.[20] Yet, the provinces have stood by the accord if one takes into account the annual allocations. 'The significance of the document, however, lies less in terms of water issues between the provinces but more in terms of national unity as it demonstrates different federating units working together under one agreement', is how Dr Hassan Abbas, an expert in hydrology and water resources, sees the document's significance.[21]

The devil, as Mustafa finds, is in the detail where the ten daily average flow allocations are allegedly, often violated, by Punjab against Sindh and sometimes by Sindh against Balochistan.[22] Further, he feels that that the problem lies because more water has been allocated to the provinces than is actually available in the system. This was done on the basis of assumption that there will be more surface storage made available on the system.[23]

---

[19] Mustafa, D. (2013). *Water Resources Management in a Vulnerable World the Hydro-Hazardscapes of Climate Change*. New York: I. B. Tauris.

[20] Ibid.

[21] Author interviewed Dr Hasn Abbas over an email.

[22] Mustafa, D. (2013). *Water Resources Management in a Vulnerable World the Hydro-Hazardscapes of Climate Change*. New York: I. B. Tauris. Also see, Mustafa, D., Gioli, G., Milan, K., & Khan, I. (2017, April 5). Contested Waters: Subnational Scale Water Conflict in Pakistan. *United States Institute of Peace: Making Peace Possible*. https://www.usip.org/publications/2017/04/contested-waters-subnational-scale-water-conflict-pakistan.

[23] Ibid.

However, he adds, the real system behaves a little differently than the civil engineers' model of the system. So the time lags that inevitably happen between the time that Sindh needs water for its earlier sowing season and the time that water can actually reach it, if IRSA allows as such, becomes a major issue. He concludes that Punjab does frequently takes more water than it is authorized and during critical times. More than the volume of waters, it is the timing of release to the lower riparian provinces which matters.

Hasan adds, as long as the Indus Delta is eroding, complaints from Sindh are justified. 'Without robust hydrological science behind policies, allocation formulas, or distribution mechanisms, the problems of shared water resources cannot be resolved, no matter how good the intentions are at the time of making these decisions'.[24]

Unfortunately, the structural and administrative mechanisms of Pakistan's water distribution systems, says Abbas are 'designed' to distribute water on 'time-slot' bases and not on 'volumetric bases'.[25] "This dilemma presents a significant scientific challenge due to which downstream users cannot see the benefits of sharing formula manifesting at their ends—not precisely because upstream users are playing foul but because the accord is not preceded by sound scientific studies, he said."[26]

Having completed 25 years last year, experts say it is one of the most significant legal documents (after the Indus Water Treaty—an agreement on sharing of waters between India and Pakistan), to protect the rights of share of the waters of the Indus Basin between the provinces. But has the time come to review or revise it?[27]

Mustafa says the document negotiated by engineers is based upon certain assumptions and habits of the mind regarding water, e.g. about volumetric flows and complete dependence upon average as a model for understanding water.

'But the accord is absolutely silent on groundwater, and the groundwater-surface water linkages–something that our water managers do not understand and don't want to talk about. The accord could be

---

[24] Author interviewed Dr Hasan Abbas over an email.

[25] Ibid.

[26] Ibid.

[27] Government of India, Ministry of Water Resources & Ganga Rejuvenation. (1960). *Indus Waters Treaty.* http://mowr.gov.in/sites/default/files/INDUS%20WATERS%20TREATY.pdf. Accessed August 12, 2018.

updated if Pakistani water managers and society had changed. They haven't so this is what you have to work with till such time that different people from different disciplines come to populate the Pakistani bureaucracy, and the Pakistani society and polity actually become democratized and multiple values of water beyond its agricultural volumetric uses become serious topics of discussion'.[28] He concludes saying the accord is fit for purpose of the society and bureaucracy it has at the moment.

Environmental lawyer, Rafay Alam Khan, on the other hand sees no significance in the piece of legislation. 'It's built on the premise that Pakistan will develop 114 MAF of storage; we haven't. So instead the accord operates on an interim arrangement that is disputed between Punjab and Sindh'.[29]

Abbas takes it on somewhat like this: 'As long as we are planning our future in the 21st century on a 170 years old gated-canal based flood-irrigation system, no treaty, accord or telemetry measurement system is going to solve our problems. While the provincial skirmishes are hovering in the tunes of 3 to 5 MAF in any given year, the annual wastage of water in irrigation system is in the tune of 50 to 60 MAF! Instead of reviewing (and improving) the accord, if we fix the wastage in irrigation sector, such accords would become irrelevant in future'.[30]

In addition, there is a plethora of documents and reports to inform policy like Planning Commission's 2014s Pillar IV (with focus on energy, water and food security) of the document called Pakistan Vision 2025; Water and Power Development Authority's Vision 2025 (2003); Asian Development Bank's 2017 report titled: A Region at Risk—The Human Dimensions of Climate Change in Asia and the Pacific and the World Bank's 2005 report titled: Pakistan's Water Economy: Running Dry.

And despite so many players and so many levels of water discourse, there is no single solution. That is because there is no innovative thinking within the sector. Though some lip-service is paid to improving the management of water, even that is only a half-serious attempt.

Experts lament that it is dominated by engineers and by the government with the former taking a technological and engineering approach of building an infrastructure to solve the country's water problem and the latter,

---

[28] Author interviewed Dr Daanish Mustafa over an email.

[29] Author interviewed Rafay Alam Khan over an email.

[30] Author interviewed Dr Hasn Abbas over an email.

Can India Turn off the Tap?

On top of that, being a lower riparian, there is always the fear that India may stop the water from flowing into Pakistan. Every few months one hears of disputes related to water between Pakistan and India which have been going on since 1948 (when India blocked the flow of water to key canals in Pakistan).

Pakistan feels India tries to intimidate Pakistan and threatens to revoke the Indus Water Treaty, a water sharing legal framework (with guidelines) between India and Pakistan signed by both the countries in the 60s. However, it is one which cannot be scrapped without mutual consent. Interestingly, this 74 pages long document with 12 articles and 8 annexures, has no expiry date.

According to the IWT the use of the eastern rivers (Sutlej, Beas, and Ravi) have been allocated to India, while Pakistan is entitled to the unrestricted use of the waters from the three western rivers—Indus, Chenab and Jhelum. India is also allowed to use the western rivers for hydropower, domestic use, specified agricultural use and non-consumptive use. It took a good nine years for the World Bank to help the two countries to sign on the dotted line.

The exchange of information exchange between the two countries regarding their use of the rivers and any issues that may arise are handled by the permanent Indus Commission of each country and differences are to be resolved by a neutral expert.

The foremost disagreement between the two countries has been construction of the Kishenganga (330 megawatts) and Ratle (850 megawatts) hydroelectric power plants, on Jhelum and Chenab rivers, respectively, being built by India. They are run-of-the-river hydropower projects that do not hold back any water, though Pakistan's objection is on the height of the gates in the dams through which waters flow downstream. Pakistan objected that both are in violation of the treaty, a claim India has rejected.

Although there are issues that the treaty does not address—like the sharing of the groundwater or factors that had not been taken into consideration back in the 50s and the 60s when the treaty was being documented, like climate change and the rising population—its proponents say the treaty is sacrosanct and cannot be reviewed at least not in the present times.[31] They say IWT is the only agreement between the warring nations that has withstood four wars or even when the relations between the two are at its lowest ebb.

Opponents, however, say the treaty has almost become irrelevant. 'If you look through another lens, you'll see a document good for the short term (what's 50-60 years for a nation state!) but one that's beginning to show its age',[32] says Alam.

In fact, in 2017, India's prime minister, Narendra Modi, taking a hardened stance after attacks on Indian Army camps in Kashmir by suspected militants, threatened scrapping it.[33] Although, after understanding of the working of the treaty, the Indian government decided against abrogating it. Instead, say some, it is finding ways of maximizing the amount of water India can use on its side.

Giving a historical perspective, Alam explains: 'Pakistan and India were nascent nation states and the World Bank was a nascent organization. The countries desperately required economic development for their people; and the Bank wanted clients. The Bank offered both countries funds to develop water infrastructure in their countries if they could agree on how to use the Indus Basin amongst themselves. If you view the treaty through these lens, it seems like a remarkable feat of "hydro-diplomacy" that has withstood the test of time'.[34] Incidentally, it is the only transboundary water document that divides; doesn't share. 'Who in their right minds divides the waters of a river system instead of sharing them?'[35]

In an exclusive interview to this writer for The Third Pole, Pakistan's former Indus Water Commissioner Jamait Ali Shah said he suspected India

---

[31] Rao, Z. (2018, October 25). Spare the Indus Water Treaty Please! *Daily Times*. https://dailytimes.com.pk/314230/spare-the-indus-water-treaty-please/.

[32] Author interviewed Dr Rafay Alam Khan over an email.

[33] *The Times of India*. (2016, September 27). "Blood and Water Can't Flow Simultaneously" PM Narendra Modi Gets Tough on Indus Treaty. Retrieved August 20, 2018 from https://timesofindia.indiatimes.com/india/Blood-and-water-cant-flow-together-PM-Narendra-Modi-gets-tough-on-Indus-treaty/articleshow/54534135.cms.

[34] Author interviewed Dr Rafay Alam Khan over an email.

[35] Ibid.

pushing Pakistan for Treaty II only on the western rivers which had been allocated to Pakistan under the IWT.[36]

Pakistan has asked the World Bank—the designated IWT mediator—to intervene but the latter has suspended all proceedings, especially on appointment of a neutral expert. Many Pakistani water experts fear India has 'lobbied aggressively and influenced' the international financial institution when the latter needs to play its part more than ever.[37]

If not outright, can the treaty be tweaked? To that Alam says: 'Who will revise this, and to what end; to secure what? No legal document produces water. It can only regulate what's already there'.[38] He further asks how does one tweak it—by including mention of climate change or by adding China or Afghanistan (also basin riparians) or by changing the dispute resolution mechanisms. 'Who in Pakistan - a lower riparian - will take the political chance that revising the treaty might compromise whatever leverage Pakistan has over India?'[39]

In addition, Alam draws attention to the current life of the treaty which revolves around the extent of India's right to use the waters of the western rivers for hydropower. 'It's basically electricity from surface water and in my opinion - and this after reading the report that says large hydropower is demonstrably unsustainable for developing economies; as well as being aware of the growth of alternate energy - India will stop building dams in the next decade. They're far too costly, and involve too much legal risk (Pakistan's legal challenges amount to costly delays) and may not be the only way to generate electricity for their growing economy. In my opinion, visionary leadership and energy policies can relegate the treaty back into the shadows (where it was – from the 1960s to the early 2000s)'.[40]

---

[36] Ebrahim, Z. T. (2017, January 6). Pakistan Not Doing Its Homework on Indus Waters Treaty, Says Former Commissioner. *thethirdpole.net*. https://www.thethirdpole.net/en/2017/01/06/pakistan-not-doing-its-homework-on-indus-waters-treaty-says-former-commissioner/.

[37] Gupta, J., & Ebrahim, Z. T. (2017, January 6). *Win Some, Lose Some, Indus Waters Treaty Continues*. https://www.thethirdpole.net/en/2017/01/06/win-some-lose-some-indus-waters-treaty-continues/

[38] Author interviewed Dr Rafay Alam Khan over an email.

[39] Ibid.

[40] Ibid.

Irrigation Efficiency

Agriculture continues to be the mainstay of Pakistan's economy. It contributes 18.9% to GDP and absorbs 42.3% of labour force.[41] The country produces seven types of agricultural products within its agricultural economy and which account for almost 60–65% of our agricultural output. They include dairy, poultry, livestock, fruit, vegetables, fisheries and flowers.

Pakistan has two cropping seasons, Kharif being the first sowing season starting from April to June and is harvested during October–December. Rice, sugarcane, cotton, maize, moong (lentils), mash (lentils), bajra (millet), and jowar (sorghum) are Kharif crops. Rabi, the second sowing season, begins in October–December and is harvested in April–May. Wheat, gram, lentil (masoor), tobacco, rapeseed, barley and mustard are Rabi crops. But Pakistan's agricultural productivity is greatly dependent upon the timely availability of water.[42]

During 2017–2018, the water available for Kharif crops in 2017 stood at 70.0 MAF showing a decrease of 2.0% over 2016 and increase of 4.3% over the normal supplies of 67.1 MAF. In Rabi season 2017–2018, the water availability stood at 24.2 MAF showing a decrease of 18.5% over 2016 and 33.5% less than the normal availability of 36.4 MAF.[43]

Globally, agriculture consumes approximately 70% of all freshwater extracted, but in Pakistan it is about 93%. Roughly 50% of irrigation needs are met by the Indus Basin Irrigation Sytem canals and 50% is extracted from the ground. According to Pakistan Bureau of Statistics (PBS) in 2013–2014 a total of 18.85 million hectares of land was irrigated through various sources including tube wells, canals and wells.

The British designed the IBIS between 1847 to 1947 to cultivate 67% of the basin area. But in the last over 70 years since the British left, Pakistan has been adding new dams, barrages, link and branch canals to the old system. Today the IBIS has three large dams, 85 small dams, 19 barrages, 12 inter-river link canals, 45 canal commands and 700,000 tube wells.[44]

---

[41] Ministry of Finance, Government of Pakistan. *Pakistan Economic Survey 2017–2018*. http://www.finance.gov.pk/survey/chapters_18/02-Agriculture.pdf.

[42] Ministry of Finance, Government of Pakistan. *Pakistan Economic Survey 2017–2018*. http://www.finance.gov.pk/survey/chapters_18/02-Agriculture.pdf.

[43] Ministry of Finance, Government of Pakistan. *Pakistan Economic Survey 2017–2018*. http://www.finance.gov.pk/survey/chapters_18/02-Agriculture.pdf.

[44] Author's conversation with Muhammad Azeem Ali Shah, regional researcher at International Water Management Institute (IWMI).

If Abbas had his way he would give a complete paradigm-shift from "supply management through mega structures to demand management" by using technologies and tweaking policies, legislation and turning around institutions and building their capacities and bringing around business models etc.[45] "It could be the game changer in the region as these interventions can potentially reduce irrigation requirements by 90%, rendering mega diversions irrelevant," he says enthusiastically.[46]

And in the last over seventy years, the cropping intensity has increased by 150% with farmers not wanting to leave any land fallow. In addition two or three crops are cultivated in a year. But the quantity of the water has remained the same. At the same time agriculture is competing with other sectors, such as industry, as well as the growing population.

According to Khalid Mohtadullah, a water policy expert and senior advisor to the International Centre for Integrated Mountain Development (ICIMOD) a regional intergovernmental learning and knowledge sharing centre, about 95% of Pakistan's freshwaters is used in agriculture sector.

Still, the country remains among those with the lowest productivity with a per acre yield dismally low. Compendium on Environment Statistics of Pakistan, published by Pakistan Bureau of Statistics in 2015 accepted the fact that the crop production in the country is lower than the world average because of inadequate availability of water at critical times during the crop growth.[47]

In fact, the Water and Power Development Authority (WAPDA), a government-owned public utility maintaining power and water in Pakistan, states that the country has the lowest productivity per unit of water i.e. 0.13 kg/m^3 in the region, compared to India at 0.39 kg/m^3 and China at 0.82 kg/m^3.[48] The same is endorsed by Anwar who notes that Pakistan continues to expand the irrigated area when it should have turned to getting more out of the area.

Earlier this year, in 2018, the Express Tribune citing the Pakistan Business Council reported that Pakistan produced 3.1 tonnes of wheat per

---

[45] Author interviewed Dr Hasan Abas over an email.

[46] Ibid.

[47] *Compendium on Environment Statistics of Pakistan, 2015.* Pakistan Bureau of Statistics, Government of Pakistan. http://www.pbs.gov.pk/sites/default/files//crops_and_climates/compendium_environment/compendium_environment_2015.pdf.

[48] Khan, M. H. (2018, January 1). Will We Be Able to Go Beyond the Rhetoric to Face the Water Challenges Ahead? *Dawn* https://www.dawn.com/news/1380112.

hectare compared to 8.1 tonnes in France; and 2.5 tonnes of cotton compared to 4.8 tonnes in China. Sugarcane—63.4 tonnes in Pakistan and 125.1 tonnes in Egypt. Rice—Pakistan 2.7 tonnes compared to 9.2 tonnes per hectare in the US. No crop has seen Pakistan at parity or even close to parity with other producers.[49]

There are many reasons contributing to low yield compared to the world average. These include wasteful irrigation methods, ill-trained farmhands and use of traditional farming techniques, poor seed varieties, inadequate land preparation and improper fertilizer application.

Experts say a fine and uniform bed is required for uniform seed germination which contributes greatly to a good harvest. Uneven seedbed, on the other hand, creates problems such as water logging and salinity and loss of nutrients and moisture, etc. And for fine and uniform seedbed, it is imperative to use a standard number of ploughings and cultivation but often farmers ignore that and continue practising traditional approaches.

About 55–92% seed sown in the country is uncertified. Scientists say recommended seed cultivar sown in a particular zone can produce optimum yield when other factors such as temperature, rainfall, wind, and humidity, etc. are also kept in view.

In addition, crop density is also not followed by farmers. With the result the plant population remains low which results in low yield. Also, the seeds sown by the farmers, often contain impurities such as sand, silt, clay, seed of other crops, weed seed and dust, etc. The seed sown by the growers may contain impurities like sand, silt, clay, seed of other crops, weed seed and dust, etc. So, on the one hand, this will result in less number of plants per unit area, on the other, generate problem of insect, pests and diseases.

Insects, pests, disease and weeds cause yield reduction up to 20% or more during pre- and post-harvest periods. Lack of quality control, high cost, adulteration, timely unavailability and lack of education and the use of faulty equipments by untrained labour are the major constraints responsible for the ineffectiveness of pesticides.

The rain-fed barani areas are heavily dependent on the seasonal June–September monsoon showers for irrigating the crops. There is a wide difference between yield of irrigated crops and barani crops. Being an arid country, the rainfall (the average annual rainfall is 291 mm) falls short of

---

[49] Hanif, U. (2018, January 24). Pakistan's Agriculture Productivity Among the Lowest in the World. *The Express Tribune*. https://tribune.com.pk/story/1616347/2-pakistans-agriculture-productivity-among-lowest-world/.

what crops would need and therefore you need irrigation to make up the difference. And with Pakistan in the throes of climate change, it will mean frequent incidences of unpredictable rains and continuous drought causing huge losses to farmers.[50]

Regular water supply is essential to sustain crop productivity. If one or two critical growth stages go without waters, it results in a significant reduction of crop production.

The answer lies in investment in capacity, in tools, in techniques in infrastructure, said Anwar.[51] To cope with water shortage, a complete reorganization of water sector institutions through mergers, economic utilization of water resources, procurement of additional storage for crops round the year, building storage to overcome droughts and to develop comprehensive water and hydro resource policy are necessary.

But, the problem says Anwar, is, that agriculture does not pay, and therefore, 'for the private sector/individual, it is not worth making any substantial investments'.

'Now you could argue "well lets raise support prices of wheat to make agriculture so much more productive and the investment worthwhile", but then we are already supporting wheat above international prices! The thing is other countries can produce wheat at a much cheaper rate than we can—perhaps because they have better climate, better rain, etc. so we cannot compete with them—no more than we can compete with China in manufacturing or India in IT!' argues Anwar.[52]

Trading Water

Water trading could be one way to resolving water tensions between provinces and, to some extent, overcoming water shortages. Taking the example of Khyber Pakhtunkhwa (KP) province, Anwar points out that with relatively less land it is easy to irrigate in the Peshawar valley, and therefore there is a surplus of water under its inter provincial water share. 'KP can either continue to seek investments in irrigated agriculture or it could simply "sell" the water to the Capital Development Authority in bulk

---

[50] *Compendium on Environment Statistics of Pakistan, 2015.* Pakistan Bureau of Statistics, Government of Pakistan. http://www.pbs.gov.pk/sites/default/files//crops_and_climates/compendium_environment/compendium_environment_2015.pdf.

[51] Author interviewed Dr Arif Anwar over an email.

[52] Ibid.

to supply Rawalpindi and Islamabad (facing water shortages) with potable water', he suggests.[53]

In turn, the KP government can use the financial resources thus mobilized to invest in human resources. 'Training more young people as, say nurses or care providers for the elderly, who can then go and work in Islamabad or Karachi looking after middle class older people and earn a decent salary. Which one do you think will raise the GDP per capita of a person from KP, growing more wheat or working in a city providing services to an affluent middle class?' he asks.[54]

Unfortunately, the water apportionment accord inhibits this trading. In fact, if you ask Anwar, he will tell you the document is so sacrosanct that policymakers to date feel very uneasy holding a conversation around it.

Greening Agriculture

However, if we still want to continue with agriculture, given water shortages, agriculture scientists say there is an urgent need for shifting towards climate smart agricultural practices.

The good news is some are in their initial stages in Pakistan. Alternative wet and drying of paddies, laser land levelling, conservation agriculture, no-till practices, biological pest management, renewable energy technologies in agricultural production systems and precision farming are already happening in pockets but need to be taken to scale and the pace increased.

Another innovation that has been put to practice is taking the pulse of the earth using satellites and models. With the help of Sustainability, Satellites, Water, and Environment (SASWE), a research group at the University of Washington, the PCRWR is now able to estimate the amount of water required by a crop at a specific location and a specific time.[55] The farmers are then sent text messages in Urdu on their cell phones about how much water their crops need that week as well as weather forecasts. This project came about because while exploring ways to improve groundwater conservation and crop yield, the PCRWR field researchers had found that farmers were over-watering their crops.

[53] Ibid.

[54] Ibid.

[55] See SASWE Research Group. http://saswe.net/.

But how do policymakers coerce farmers to move away from age-old irrigation methods. Abbas has a simple solution. 'Devise mechanisms that reward efficiency in the irrigation sector and promote emerging efficient technologies so that they become accessible to farms'.[56] And if that happens, 'outdated technologies' of large dams and mega diversions could be systematically phased out.[57]

Spoilt Produce

Despite so much production, however, what is most unfortunate is that as much as 50% of the output is wasted because of the unavailability of a cold chain (integrated system of storage, transportation and distribution of perishable and temperature-sensitive commodities), logistics in the form of inefficient crop extraction and poor processing.[58]

GROUNDWATER

NASA's researchers found that of the planet's 37 largest aquifers studied between 2003 and 2013 the Indus River Basin aquifer is the second-most overstressed among them. It is depleting without receiving waters to recharge the basin. This basin is also on the World Water Resource Institute's water stress index.[59]

Take the case of the eastern city of Lahore where the residents are supplied with groundwater. A 2014 study by the WWF-Pakistan found that the residents of Lahore, over six million, were supplied by the Water and Sanitation Authority (WASA) over six million population through some 484 tube wells. Located at different places, their depth varied between 150 to 200 m.[60] Since with time, water demand has increased from 180 litres per capita per day (lpcd) in 1967 to 274 lpcd in 2013, so did abstraction of water. Today it stands at about 2.2 million cubic metres per day

---

[56] Author interviewed Dr Hasan Abas over an email.

[57] Author interviewed Dr Hasan Abas over an email.

[58] Shaikh, A. (2018). Unlocking the Cold Chain Potential. *Aurora*. https://aurora.dawn.com/news/1143185.

[59] Study: Third of Big Groundwater Basins in Distress. (2015, June 17). *NASA*. https://www.nasa.gov/jpl/grace/study-third-of-big-groundwater-basins-in-distress.

[60] Situation Analysis of the Water Resources of Lahore Establishing a Case for Water Stewardship. *WWF Pakistan*. http://www.wwfpak.org/wsp/pdf/SAWRL.pdf.

(MCM/day) from those tubewells, running 14-18 h per day with water distributed from source to households through a network of 7700 km long water supply lines and 600,000 connections. Compared to 78% of households in the WASA serving area that are connected to the piped water, just 50% of those living in non-WASA areas enjoy this facility.[61] The remaining 50% draw water from hand pumps, go to public water stand posts use suction pumps and pump up the groundwater directly.

In the absence of any municipal water act or water-right law, groundwater is extracted without fear or remorse by private housing schemes and industry. In Lahore, private housing societies supply water to its residents by pumping up to 0.37 MCM/day. In areas where there is no water supply network, the estimated extraction is 0.35 MCM/day. In all the total groundwater extracted is approximately 0.71 MCM/day.[62]

In villages surrounding Lahore, the water is supplied by the Public Health Engineering Department (PHED). Of the 16 rural water supply schemes that are supposed to provide water to these areas, 13 are nonfunctional due to non-payment of electricity bills. In conclusion, the total domestic water use in Lahore is estimated at 3.79 MCM/day (1384 MCM/year) and groundwater is being used faster than it is replenished.[63]

With all sectors, including for domestic use, industry and agriculture, relying on groundwater, the total groundwater discharge is 7.17 MCM/day (2619 MCM/year). Agricultural sector uses some surface water resources. The domestic sector consumes the largest share (53%) of the groundwater, followed by the[64] industrial sector consuming 13%, the agriculture using 24% and the remaining 10% consumed by the institutional sector.[65]

With groundwater recharge at 6.50 MCM/day (2372 MCM/year) and the discharge at 7.17 MVM/day, the difference of 0.67 MCM/day (247 MCM/year) means an equivalent of 55 cm (0.55 m) per year drop in aquifer levels. This excessive pumping also means the water table depth has gone below 40 m, and projected to drop to below 70 m by 2025 in most areas.[66]

---

[61] Ibid.
[62] Ibid.
[63] Ibid.
[64] Ibid.
[65] Ibid.
[66] Ibid.

Bottled Water Companies in Judiciary's Cross Hairs

While the citizens are clamouring for water, it seems bottling companies have an endless supply to fill up their bottles and that too almost for free. The Chief Justice of Pakistan recently took notice of this and directed all provinces to impose a tax of one rupee per litre on companies bottling and selling water.

While the consumer of bottled water pays for the groundwater in the form of—cost of drawing water from the source, the raw material used in the manufacture of the plastic bottle, energy used in treatment of water, energy used in cleaning, filling, labelling, refrigerating and, marketing, and transportation cost to retailers and then to the consumer, the company pays nothing for using that water to run its business.[67]

Free for All

Pakistan, unfortunately does not have any law specific to groundwater abstraction. But there are references and sections pertaining to the use of groundwater in different laws and acts. The National Water Policy 2018, or the earlier Canal and Drainage Act 1873, the Provincial Irrigation and Drainage Authority Act 1997 or the Punjab Soil Reclamation Act 1952 have scarce mentions with words like 'groundwater', 'sub-soil' or 'underground resources'. Prime Minister Imran Khan has directed the Ministry of Water Resources and Planning Division to coordinate with provincial governments and work out a comprehensive plan along with a legislative framework for ensuring maximum utilization of the surface water and reversing the existing trend of unabated pumping of groundwater.

In 2017 the Punjab government's Irrigation Department had drafted the Punjab Groundwater Protection, Regulation, and Development Act (2017) to regulate groundwater extraction across Punjab province and had even started consulting with various stakeholders. When it will get a shot in the arm is anyone's guess.

But there no limit to how much the bottling companies can extract, they also do not pay for what they pump out. Instead of paying against their volumetric extraction they pay based on the size of land they occupy. It is a fixed monthly commercial rate based on the size of land prescribed under

---

[67] Ebrahim, Z. T. (2018, October 21). The Real Cost of Bottled Water. *Dawn*. https://www.dawn.com/news/1440240.

provincial development laws, say like the Punjab Development of Cities (Amendment) Act 2014, in case of Punjab.

Bottling companies say the industry uses approximately 0.001% of the 2% allocated for industrial usage—spread among the approximately 100 bottled water brands present in Pakistan.

This is corroborated by Anwar. In terms of volumes, he pointed out the amount of water abstracted for bottling is minuscule compared to volumes used for agriculture, particularly surface irrigation. He emphasized the need for Pakistan to move towards charging all sectors for using water be it agriculture, municipality or the industry.[68]

'Put a price on it!' said Anwar. 'When I say charge for water, I also include surface water users and the charge would include water as well as its delivery or abiana'.[69]

Urban Water Supply and Sanitation Issue

According to the World Bank, Pakistan has made substantial improvement in access to water and sanitation. In addition there has been seen a decline in open defecation. Today, 90% of homes have access to 'improved' water and some 73% enjoy 'improved' toilet facilities.[70]

Still, just 25% of households, a majority living in cities, have access to piped water. In contrast, some 60% made their own arrangements by installing hand or motorized pumps. Another 11% continued to rely on completely unprotected sources.[71]

As for sanitation, only 22% of households have access to toilets that are connected to underground sewer systems. Most of these are in cities. For the rest it is septic tanks. This is mainly in the better off districts of Punjab and KP. In Sindh and Balochistan, more than 75% of villagers rely on very basic latrines.[72] Open defection, use of poor quality sanitation facilities and water from unimproved sources are still common features. Little wonder this issue is haunting Prime Minister Imran Khan who has

---

[68] Author interviewed Dr Arif Anwar over an email.

[69] Author interviewed Dr Arif Anwar over an email.

[70] The World Bank Report. (2019). *Pakistan: Getting More from Water.* https:// openknowledge.worldbank.org/handle/10986/31160.

[71] Ibid.

[72] Ibid.

directed the Ministry of Water Resources and Planning Division to form a comprehensive plan on urban water schemes for all the major cities on a priority basis to overcome water scarcity.

The World Bank has pointed out that the state has failed in providing water and sanitation services. Provision of piped water has declined becoming more unreliable in terms of hours of service, particularly in Sindh and Balochistan, leading to people turning to private water tanker suppliers. People have also installed motorized hand pumps in places where groundwater extraction is economically feasible.[73]

Solutions

The World Bank offered the following solutions[74]:

1. Address the institutional and implementation challenges

   - Reduce the overlap of responsibilities and narrow the coordination gap.
   - Demarcate responsibilities of each water and sanitation department.
   - Every province to have its own policy to improve coordination among departments.
   - Replace vertical grants with sector-specific funds that are conditional on performance.
   - Operations and maintenance (O&M) a key part of institutional accountability.
   - Mobilize communities to enhance accountability, not to provide O&M.

2. Improve technical capacity

   - Develop capacity of public sector staff.
   - Involve the private sector to help fill the technical capacity gap.

---

[73] Ibid.

[74] The World Bank Report. (2019). *Pakistan: Getting More from Water.* https://openknowledge.worldbank.org/handle/10986/31160.

3. Monitor the sector

- Have a sector management information system (MIS).
- Improve the quality of survey data collected for monitoring SDGs related to water and sanitation by the national and provincial bureaus of statistics.
- Ensure definitional consistency between MIS and survey data.

4. Improve water quality on an urgent basis

- Invest in point-of-use water treatment (chlorination or other method), with the use of subsidies.
- Target 100% piped water supply in the long run, with metering and realistic tariffs to cover O&M.
- Develop a regulatory framework for groundwater.

5. Improve sanitation infrastructure

- Invest in faecal waste management (treatment of sludge and wastewater).
- Regulate drainage systems and septic tank designs and enforce a safe distance from water sources.
- Create a regulatory body to set and enforce standards for both public and private providers.

6. Rationalize the allocation of district budgets for WASH

- Reallocate existing spending toward districts with the greatest needs.
- Use multisectoral planning to maximize the benefits from investments.
- Establish a clear allocation system for sanitation-related schemes in the budgeting process.
- Budget for O&M upfront.

POLLUTION AND CONTAMINATION

Pakistan's rivers are laced with raw sewage and untreated industrial effluents. A recent study (and published earlier this year in Chemosphere), jointly conducted by Pakistan and China, found three of Pakistan's major rivers—Chenab, Indus and Kabul—contaminated with potentially toxic elements (PTE). Researchers collected water, sediment and fish samples from the sites along the rivers and analysed them at the Institute of Tibetan Plateau Research (ITPR), Beijing.[75]

The samples were contaminated 'beyond safe limits' with arsenic, lead, cadmium, chromium, nickel, copper, cobalt, manganese and zinc according to the researchers.[76]

Since early this year, the Punjab Food Authority has been destroying standing crops on thousands of acres that had been watered with untreated industrial effluent and sewage.[77] In the future, the farmers are only be allowed non-edible crops like bamboos, flowers, and indoor plants.[78]

But surface water alone is not the only one contaminated. The groundwater is as polluted—by industrial and municipal effluent.

Last year, a research published in Science Advances analysing data from nearly 1200 groundwater samples from across Pakistan stated that up to 60 million people were at risk from the deadly chemical arsenic. The World Health Organisation has established a level of 10 micrograms per litre as the permissible concentration in drinking water, but Pakistan government says that 50 micrograms per litre is acceptable. The study said 'very high concentrations, above 200 micrograms/liter, are found mainly in the south'[79] and warned that regular consumption of water containing high concentrations of arsenic may lead to skin disorders, lung cancer, and cardiovascular diseases.

Contamination is particularly worrying in this case as Pakistan is unusually dependant on a single, vast underground natural reservoir known as the Indus basin aquifer. The aquifer covers an area of 160,000 km^2—making it slightly larger than England—and spans Pakistan's border with India.

---

[75] Nawab, J., Khan, S., & Xiaping, W. (2018, July). Ecological and Health Risk Assessment of Potentially Toxic Elements in the Major Rivers of Pakistan: General Population vs. Fishermen. *Chemosphere, 202*, 154–164.

[76] Ibid.

[77] Ahmed, S. I. (2018, January 11). Pakistan Launches Drive Against Crops Grown with Wastewater. *thethirdpole.net*. https://www.thethirdpole.net/en/2018/01/11/pakistan-launches-drive-against-crops-grown-with-wastewater/.

[78] Vegetable Growers Using Untreated Industrial Wastewater Face PFA Crackdown. (2018, November 5). *The News*. https://www.thenews.com.pk/print/389692.

[79] Podgorski, J. E., Eqani, S. A. M. A. S., Ullah, R., Shen, H., & Berg, M. (2017, August 23). Extensive Arsenic Contamination in High-pH Unconfined Aquifers in the Indus Valley. *Journal of Sciences, 3*(8). http://advances.sciencemag.org/content/3/8/e1700935.full. Accessed September 30, 2018.

The transboundary Indus river basin has a total area of 1.12 million square km distributed between Pakistan (47%), India (39%), China (8%) and Afghanistan (6%).[80] The Indus river basin stretches from the Himalayan mountains in the north to the dry alluvial plains of Sindh province in Pakistan in the south and finally flows out into the Arabian Sea. In Pakistan, the Indus river basin covers around 520,000 km^2, or 65% of the territory, comprising the whole of the provinces of Punjab and Khyber Pakhtunkhwa and most of the territory of Sindh province and the eastern part of Balochistan.[81]

Contaminated water has also entered the homes of residents in urban centres as well. Take the case of Sindh. A 2016 survey conducted by the Sindh Water Commission set up by the Supreme Court of Pakistan found 83.5pc of water in 14 out of 29 districts in Sindh is unsafe for drinking. The water samples were taken from various sources including rivers, canals, reverse osmosis plants, water supply schemes.

Of all the urban centres, Karachi had the highest water supply contamination score in the report. Of the 114 samples collected, 104 (88.1%) were found to have presence of coliform bacteria beyond the World Health Organisation values (0/100 ml cfu) and 40 (33.4%) had faecal contamination (presence of E. coli). The overall data showed that 107 (90.7%) samples collected from various places in Karachi were unsafe for drinking purposes.[82]

Waterboard officials conceded contamination but laid part of the blame on the residents, who steal water. The water, the board said, is stolen from the underground mains by puncturing them. Karachi has 10,000 kilometres of both old iron pipelines and new ones made of high-density polyethylene plastic pipes snaking through the city. The sewerage often gets mixed into the water through these punctures. In addition, there is a need to rehabilitate and replace the entire distribution network, a gargantuan and expensive task which may require two to three years.

In 2016, the requirement for Karachi was 1100 million gallons per day (mgd) for a burgeoning population of 25 million, over half of whom live in

---

[80] Food and Agriculture Organization of the United Nations. (2011). *Indus Basin*. http://www.fao.org/nr/Water/aquastat/basins/indus/index.stm.

[81] Ibid.

[82] Ebrahim, Z. T. (2017, July 28). Revealed: 91% of Karachi's Water Unfit to Drink. *thethirdpole.net*. https://www.thethirdpole.net/en/2017/07/28/91-of-karachis-water-unfit-to-drink/.

squatter colonies, but the Karachi Water and Sanitation Board (responsible for the supply of 90% of water in the metropolis) was able to supply only between 450 to 480 mgd.[83] In addition, KWSB also supplies water to nearby areas of Karachi like Dhabeji, Ghaggar and Gharo.

DAMS

Dams—big dams, small dams and the ones in between—dams of all sizes and kinds need to be built on a war footing because failing to do so would be disastrous for Pakistan. For years now, water experts have been clamouring for new storages against a backdrop of rapid urbanization, population growth, food insecurity and growing water demand from industry. The clamour increases every time the country is swallowed by a deluge due to heavy downpour. Pakistan has a total of 155 dams compared to the 5102 in India and can store water for only 30 days. The international standard is 120 days.

Mohtadullah has long been calling for the need for storages, in the form of field storages (ponds), small dams, big dams, all kinds of dams and everywhere, he says. 'Every drop we store is a drop saved', he emphasizes.[84] Then if such shocks arise (as when India threatens to block the water), Pakistan is better prepared to absorb them.

And there are other means of conserving water too. According to WAPDA chairman, Lt Gen Muzammil Hussain, about 12 MAF of water can easily be conserved through efficient use, another 11 MAF can be saved by provinces through better managing the demand and a further 12 MAF can be saved through lining of canals, and 35 MAF can be made available through construction of dams.[85]

No major dams have been constructed in Pakistan since the Tarbela dam in 1976. Along with Mangla dam, the two major reservoirs in the Indus basin store only 14 million acre feet (MAF) of the 145 MAF that flows through Pakistan annually.

Today, leading this the call is none other than the Chief Justice of Pakistan, Mian Saquib Nisar who wants to "build a dam on every inch of

[83] Ibid.

[84] Author interviewed Dr Khalid Mohtadullah over an email.

[85] Kiani, K. (2018, June 7). Call to Make Indus Waters Treaty Part of Foreign Policy. *Dawn*. https://www.dawn.com/news/1412565.

Indus".[86] He is right now concentrating on collecting funds—Rs 1500 billion—4500-megawatt Diamer-Bhasha Dam and the 2000 megawatt[87] Dasu Hydropower Project on the river, to avert the looming water crisis.[88] The Supreme Court of Pakistan has a website showing an update on the funds gathered.[89]

These two are part of what is called the Indus Cascade (five dams in a row on the Indus) with assistance from China. Costing USD 50 billion, for 22,320 MW of hydel power (Diamer-Bhasha 4500 MW; Patan 2400 MW; Thakot 4000 MW; Bunji 7100 MW; Dasu 4320 MW). In addition to the huge cost, the hydel project is in the middle of seismically active mountains at the junction of three tectonic plates, say scientists.[90]

For years now, experts have been saying big dams are risky projects with cost overruns, debt and inflation since these require huge investment.

In 2014, Dr Atif Ansar of Blavatnik School of Government, who co-authored a research with Professor Bent Flyvberg (Oxford University's Saïd Business School), based on the most extensive dataset of its kind, came to the conclusion that the construction costs of large dams are, on average + 90% higher than their budgets at the time of approval.

Closer to home, the actual cost of Tarbela dam, most of which was borrowed from external sources, amounted to 23% of the increase in Pakistan's external public debt stock between 1968–1984, Ansar and Flyvberg had written in The Guardian.[91] And yet, Pakistan continues to show amnesiac

---

[86] We Will Build Dams on Every Inch of Indus: CJP. (2018, November 26). *The Express Tribune.* https://tribune.com.pk/story/1854481/1-will-build-dams-every-inch-indus-cjp/.

[87] Nawaz Approves Allocation for Bhasha, Dasu Dams, Motorways. (2014, May 23). *Dawn.* https://www.dawn.com/news/1108169.

[88] The World Bank. (2014, June 10). *World Bank Approves Dasu Hydropower Stage I Project.* http://www.worldbank.org/en/news/press-release/2014/06/10/world-bank-approves-dasu-hydropower-stage-i-project.

[89] Fund Raising Status for the Supreme Court of Pakistan and the Prime Minister of Pakistan Diamer-Bhasha and Mohmand Dams Fund. (2019, April 5). *Supreme Court of Pakistan.* http://www.supremecourt.gov.pk/web/page.asp?id=2757.

[90] Gupta, J. (2018, May 22). Indus Cascade a Himalayan Blunder. *Thethirdpole.net.* https://www.thethirdpole.net/en/2017/05/22/indus-cascade-a-himalayan-blunder/. Accessed May 12, 2018.

[91] Flybjerg, B., & Ansar, A. (2014, April 7). Hydroelectric Dams Are Doing More Harm Than Good to Emerging Economies. *The Guardian.* https://www.theguardian.com/sustainable-business/hydroelectric-dams-emerging-economies-oxford-research.

behaviour again with the new project that is being hailed as Pakistan's only chance to save its citizens from dying of thirst.

In addition, they may cause displacement of local communities and loss of rich biodiversity. There are others who fear big dams could become easy targets for militants, as the Mosul dam recently became in Iraq.[92]

Abbas explains that structural interventions do not create water, only store and/or divert it to where it did not exist, while simultaneously depriving it from where it existed earlier. 'The negative environmental consequences, social impacts and economic externalities of large dams and diversions remain obscure because they are usually offset in time and space'.[93]

The Oxford University study based on data from 245 large dams in 65 different countries not only proves that mega dams are not economically viable, it says they come at great human and ecological cost. Little independent research on the viability of large dams has been carried out since the findings of the World Commission on Dams in 2000. And by 2000, the number of large dams had climbed to more than 47,000, and an additional 800,000 smaller dams blocked the flow of the world's rivers.

It also claims dam building has burdened countries with large amounts of debt. But cost aside, mega dams take a long time to build—about eight to ten years, even more. With the result they are ineffective in resolving urgent power crises or benefit the politicians in winning votes. Instead, the research argued that 'smaller, more flexible hydroelectric projects' that are built faster should be the preferred choice.

Unfortunately, after a decade-long 'lull', emerging economies of Brazil, China, Ethiopia and Pakistan are again in a rush to build mega dams on an unprecedented scale, claims the report.

'The problem is even when we talk of dams - say Mohmand dam, we immediately want to put that water to more irrigated agriculture (17,000 acres or so)', says Anwar[94] and goes on: 'So how does that help with the water woes of Karachi and Islamabad where incidentally the real growth in GDP and prosperity is happening?'[95]

---

[92] Miner, A. (2014, August 18). *Mosul Dam: Why the Battle for Water Matters in Iraq.* https://www.bbc.com/news/world-middle-east-28772478.

[93] Author interviewed Dr Hasan Abbas over an email.

[94] Author interviewed Dr Arif Anwar over an email.

[95] Ibid.

CONCLUSION

As discussed in this chapter, Pakistan is facing severe water stress which has its impact on the country's agriculture dependent economy and individuals. Politically, the decline in water availability also escalates the interprovincial water disputes in Pakistan. Often the four other provinces allege that by being an upper riparian Punjab does not release adequate amount of waters for others use. In 1991 an accord was signed and institutions were set up for appropriate distribution of waters among the provinces, however, problems remain. To address the country's water woes the government is taking demand and supply sides measures whose effectiveness on the ground has to be observed.

Domestic Water Stress, Transboundary Tensions and Disputes

Amit Ranjan

Water situation across the world is turning dangerous. It is becoming one of the most sought after natural resources because of its increasing demand and declining availability. The co-riparian states and regions are competing to get access to more of the transboundary waters so that their domestic needs can be satisfactorily addressed. On the precarious water situation, United Nations World Water Report, 2018, says that, if it remains business as usual, large parts of the world may face severe water scarcity by 2050.[1] This report defines a region as 'water scarce' on the basis of total water withdrawal limit for human use. If the withdrawal is between 20 and 40% of the total available renewal surface waters the region can be categorized as 'water scarce' and if the withdrawal exceeds 40% than it can be called as 'severe water scarcity' area.[2] Many of such regions, are in Himalayan South Asia and China. Prime reasons for the growing scarcity are the phenomenon of climate change and growing population which are constantly increasing

---

[1] United Nations Water Development Report. (2018). http://unesdoc.unesco.org/images/0026/002614/261424e.pdf. Accessed May 12, 2018, p. 12.
[2] Ibid.

A. Ranjan (✉)
Institute of South Asian Studies, National University of Singapore, Singapore

© The Author(s) 2020
A. Ranjan (ed.), *Water Issues in Himalayan South Asia*,
https://doi.org/10.1007/978-981-32-9614-5_8

183

gaps between demand and supply of waters. The rising population needs industrial goods and waters for domestic consumption.

Himalayan South Asia, plus China, is one of the populous regions in the world. At present, China's population is about 1.42 billion people, India has about 1.35 billion people, Pakistan's population as more than 216 million, Bangladesh has population of about 163 million, Nepal has about 28.61 million people, and Bhutan's population is 763,092. In coming years this is likely to increase further which will simultaneously increase consumption of food, industrial goods and demand of water for domestic consumption.

In addition to increasing demand-supply gap, a large quantity of waters are being wasted in agriculture activities because of lack of technology. There is also an issue of pollution of surface and ground waters. Pollution of waters is affecting the health of a large number of people by forcing them to consume infected and polluted waters or foods grown by using such waters. This chapter takes stock of water situation in the respective countries and, argues that the growing water stress in respective countries may escalate their transboundary water disputes with co-riparian states.

GROWING WATER STRESS IN HIMALAYAN SOUTH ASIA AND CHINA

India is home to about 4% of world's water resources and more than 17% of the global population. This distribution ratio makes it already a water stress country.[3] Unfortunately, the population is growing every year while supply of waters remain, more or less, constant. Highlighting the imminent stress in 2018, for the first time in June 2018, India's National Institution for Transforming India (NITI Aayog), a government of India think tank which has replaced the planning commission in 2014 published a report titled

---

[3] Aggarwal, M. Will a New Water Ministry Solve India's Impending Water Crisis? *The Wire.* https://thewire.in/environment/water-stressed-india-management-rivers.

DOMESTIC WATER STRESS, TRANSBOUNDARY TENSIONS AND DISPUTES   185

Composite Water Management Index : A Tool for Water Management.[4] According to the report[5]

600 million Indians face high to extreme water stress about two lakh people die every year due to inadequate access to safe water. The crisis is only going to get worse. By 2030, the country's water demand is projected to be twice the available supply, implying severe water scarcity for hundreds of millions of people and an eventual 6% loss in the country's GDP [Gross Domestic Product]. As per the report of National Commission for Integrated Water Resource Development of MoWR [Ministry of Water Resources], the water requirement by 2050 in high use scenario is likely to be a milder 1,180 BCM [Billion Cubic Metre], whereas the present-day availability is 695 BCM. The total availability of water possible in country is still lower than this projected demand, at 1,137 BCM. Thus, there is an imminent need to deepen our understanding of our water resources and usage and put in place interventions that make our water use efficient and sustainable.

To address the concern of the growing water scarcity in India, during his election speeches in 2019, Indian Prime Minister Narendra Modi said that there will be a dedicated ministry called Jal Shakti (Water Power) to look at the water-related issues such as availability of clean waters, ensuring waters for irrigation etc.[6] After coming into power for the second consecutive term in May 2019, Modi kept his electoral promise and created Jal Shakti Ministry by reorganizing Ministry of Water Resources, River Development and Ganga Rejuvenation. Ministry of Drinking Water and Sanitation has also been added to the newly created ministry. Gajendra Singh Shekhawat was sworn in as the cabinet minister on 30 May 2019 to head Ministry of Jal Shakti.[7]

---

[4] NITI Aayog, Government of India. *Composite Water Management Index: A Tool for Water Management*, p. 15. http://niti.gov.in/writereaddata/files/document_publication/2018-05-18-Water-Index-Report_vS8-compressed.pdf. Accessed June 25, 2018.

[5] Ibid.

[6] "Will Create Jal Shakti Ministry To Fight Water Scarcity": PM Modi. *NDTV*, 3 May 2019. https://www.ndtv.com/india-news/elections-2019-narendra-modi-says-will-create-jal-shakti-ministry-to-fight-water-scarcity-2032593.

[7] As Promised, Modi Forms Jal Shakti Ministry. (2019, May 31). *ANI News*. https://www.aninews.in/news/national/politics/as-promised-modi-forms-jal-shakti-ministry20190531141609/.

In Pakistan, the preamble to the National Water Policy, 2018 states that 'With rapidly growing population, Pakistan is heading towards a situation of water shortage and by corollary, a threat of food insecurity. Per capita surface water availability has declined from 5260 cubic metres per year in 1951 to around 1000 cubic metres in 2016. This quantity is likely to further drop to about 860 cubic metres by 2025 marking our (Pakistan's) transition from a "*water stressed*" to a "*water scarce*" country'.[8] The availability issue has been further aggravated because of its high dependence on the upper riparian—India—with which it does not share good relationship. Also, due to its sliding economic situation, Pakistan does not have enough fund to build dams to manage its waters. In many cases it has taken loan from other countries such as China to build dams. The economic situation of Pakistan forced the government to go for crowdfunding to construct two dams-Diamer Bhasha and Mohmand.

Besides declining of water availability, Pakistan also suffers from water pollution which makes a large quantity of waters unfit for consumption. Both surface and groundwaters have been polluted because of release of a large amount of untreated industrial waters and wastes into the waterbodies.[9]

In 1999, Bangladesh adopted its first National Water Policy which states that 'Water resources management in Bangladesh faces immense challenge for resolving many diverse problems and issues. The most critical of these are alternating flood and water scarcity during the wet and the dry seasons, ever-expanding water needs of a growing economy and population, and massive river sedimentation and bank erosion. There is a growing need for providing total water quality management (checking salinity, deterioration of surface water and groundwater quality, and water pollution), and maintenance of the eco-system. There is also an urgency to satisfy multi-sector water needs with limited resources, promote efficient and socially responsible water use,

---

[8] Ministry of Water Resources, Government of Pakistan. *National Water Policy 2018*, p. 1. http://www.ffc.gov.pk/download/AFR/National%20Water%20Policy%20-April%202018%20FINAL.pdf. Accessed August 12, 2018.

[9] Ghani, A., & Ebrahim, Z. T. (2019, May 20). The Problems Caused by Mishandled Industrial Waste. *Herald*. https://herald.dawn.com/news/1398877.

DOMESTIC WATER STRESS, TRANSBOUNDARY TENSIONS AND DISPUTES **187**

delineate public and private responsibilities, and decentralize state activities where appropriate'.[10]

Although Bangladesh is fairly rich in terms of water availability, the per capita availability is declining due to rise in demand. It is projected that in 2030 there will be an increase in about 109% in industrial water demand, domestic water demand by 75%, and agricultural water demand by 43%.[11] Also, Bangladesh has little control over its water resources. Country's major rivers are transboundary—54 of them come from India and 3 from Myanmar.[12] More than it, almost all major rivers of the country are highly polluted and even the river beds are facing encroachments due to growing urbanisation. In May 2019, the Dhaka High Court found that a number of powerful individuals, businesses and, even, government offices were engaged in such encroachment activities. Delivering its judgement to save the rivers, the Court ordered the State to act as the trustee of all rivers, hills, sea beaches, forests, canals and beels (lake like wetland) and other waterbodies. In its order the Court also said that the National River Protection Commission will be a main body entrusted with a responsibility to protect them.

Even a large quantity of the supplied waters for domestic consumption in Bangladesh are not safe. On 16 May 2019 in its report submitted to the High Court, Dhaka Water Supply and Sewerage Authority accepted that the water supplied by it to 57 areas of its 10 zones are polluted.[13] A high quantity of arsenic is found in Bangladesh's groundwaters. According to World Health Organization's study, the presence of arsenic in ground water affects an estimated population of around 30–35 million of Bangladeshis. It is estimated that use and exposure to arsenic contaminated water is a reason for 1 out of every 5 deaths in Bangladesh.[14]

[10] Ministry of Water Resources, Government of Bangladesh. (1999). https://mowr.portal.gov.bd/sites/default/files/files/mowr.portal.gov.bd/files/32e67290_f24e_4407_9381_166357695653/National%20Water%20Policy%20(English).pdf.

[11] *BANGLADESH WATER SECTOR NETWORK STUDY Final Report*. Prepared by Light Castle Partners, 31 October 2018. https://www.netherlandswaterpartnership.com/sites/nwp_corp/files/2019-01/bangladesh_water_sector_network_studyreport.pdf.

[12] Ibid.

[13] Wasa Water Polluted in 57 Areas of Dhaka. (2019, May 16). *The Daily Star*. https://www.thedailystar.net/country/water-pollution-in-dhaka-wasa-water-polluted-in-57-areas-1744423.

[14] Hedrick, S. *Water in Crisis: Spotlight on Bangladesh*. The Water Project. https://thewaterproject.org/water-crisis/water-in-crisis-bangladesh.

Afghanistan is also a water-scarce country. According to 2018 report of the Ministry of Energy and Waters, total volume of waters available until few years was 76 billion cubic metres but has dropped by 10 billion cubic metres in recent times.[15] Due to the burden on underground waters and ineffective water policy, the quantity of underground waters have declined to 17 billion cubic metres from 18 billion cubic metres in the recent past.[16] It is believed that if the situation remains the same, by 2030 Afghanistan will face a lack of water in its river basins and will also experience shortage in underground water.[17] Precipitation and snowfall adds 57 billion cubic metres of water annually but due to lack of proper management only between 30 and 35% of this water are being retained in the country. As a result, about 68% of Afghan population does not have access to clean water and about 80% faces water shortage.[18]

China with a population of about 1.42 billion (2018) is another water-thirsty neighbour of India. The declining availability of waters has decreased the per capita availability of water in China to around 2100 cubic metres.[19] Country's water resources are unevenly distributed where north has scarcity but south possess about 80% of the country's waters.[20] Also a large quantity of available waters, as shown by a study by Greenpeace International in 2017 are polluted. The study found that 39.9% of water in Beijing, 65.9% of waters in Tianjin and 30.2% of waters in Hebei are not usable for agriculture or industrial purposes because of pollution.[21]

---

[15] Akbari, M. Z. (2018, April 4). Water Crisis in Afghanistan. *Daily Outlook Afghanistan.* http://outlookafghanistan.net/topics.php?post_id=20584. Accessed March 12, 2019.

[16] Ibid.

[17] Ibid.

[18] Ibid.

[19] Ministry of Water Resources, People's Republic of China. *WATER RESOURCES IN CHINA.* http://www.mwr.gov.cn/english/mainsubjects/201604/P020160406508110938538.pdf. Accessed April 12, 2018.

[20] Ministry of Water Resources, People's Republic of China. *WATER RESOURCES IN CHINA.* http://www.mwr.gov.cn/english/mainsubjects/201604/P020160406508110938538.pdf. Accessed April 12, 2018.

[21] Greenpeace. (2017, June 1). *Nearly Half of Chinese Provinces Miss Water Targets, 85% of Shanghai's River Water Not Fit for Human Contact.* http://www.greenpeace.org/eastasia/press/releases/toxics/2017/Nearly-half-of-Chinese-provinces-miss-water-targets-85-of-Shanghais-river-water-not-fit-for-human-contact/. Accessed April 21, 2018.

DOMESTIC WATER STRESS, TRANSBOUNDARY TENSIONS AND DISPUTES   189

Bhutan which is one of the water-rich countries in the world receives around 70,576 million cubic metres of waters every year.[22] This makes the per capita availability of water in the country to more than 100,000 cubic metre.[23] Nevertheless, there is a sign of imminent scarcity in the parts of Bhutan. In its study, carried out in 2016, the National Environment Commission, Royal Government of Bhutan found that some of the districts of the country like Thimphu, Haa, and Zhemgang may experience water shortage[24] by the year 2030.[25] Furthermore, the study confirmed that due to impact of the climate change, few of the important water sources in the country have started drying up.[26]

Another water-rich country, Nepal faces a number of challenges to accrue benefits from its available resource. According to Water Resources Strategy document of 2002:

> Nepal faces a number of physical and human challenges in recognizing benefits associated with water resources development. The country's rugged topography, young geology and monsoon climate all combine to produce high rates of runoff, erosion and sedimentation.....Increasing population pressure and demand for agricultural land often conflict with plans for protection of the natural environment. In urban areas, wastewater, solid wastes and air pollution have seriously degraded living conditions. Poverty and environmental degradation are closely interrelated in Nepal.[27]

---

[22] Asian Development Bank. (2016). *Water Securing Bhutan's Future*, p. 75. https://www.adb.org/sites/default/files/publication/190540/water-bhutan-future.pdf. Accessed April 2, 2018.

[23] Royal Government of Bhutan, National Environment Commission. (2014). *Bhutan Water Vision and Bhutan Water Policy*. http://www.nec.gov.bt/nec1/wp-content/uploads/2014/04/Bhutan-Water-Policy-Eng.pdf. Accessed April 2, 2018.

[24] Availability of water of 1700 cubic meters per person in country makes it water stressed. Lesser than it, availability of 1000 cubic meters of water per person represents a state of "water scarcity". Below 500 cubic meters per person is termed as the state of an "absolute scarcity". Human Development Report. (2006). *Beyond Scarcity: Power, Poverty and the Global Water Crisis* (p. 135). New York: United Nations Development Programme. http://hdr.undp.org/sites/default/files/reports/267/hdr06-complete.pdf. Accessed April 20, 2018.

[25] Royal Government of Bhutan, National Environment Commission. *National Integrated Water Resources Management Plan 2016*, p. 33. http://www.nec.gov.bt/nec1/wp-content/uploads/2016/03/Draft-Final-NIWRMP.pdf. Accessed May 21, 2018.

[26] Ibid.

[27] Government of Nepal. (2002). *Water Resources Strategy, Nepal*, p. ii. https://www.moen.gov.np/pdf_files/water_resources_strategy.pdf.

Further, underlining the pollution of available waters, report of the Department of Water Supply and Sewerage in Nepal finds, even though an estimated 80% of the total population has access to drinking water, it is not safe. The marginalize group has limited or almost no access to safe drinking waters. Many people have to cover a long distance to get waters.[28] Many of the sources of the water in Nepal are either getting dry or are too much polluted to use for any purpose. They are causing concerns for the local population in the river bed areas.[29]

Transboundary Waters Tensions and Disputes

As there are growing water stress in the respective Himalayan South Asian countries, they are competing against each other over transboundary rivers waters to meet their domestic demands. Most of the riparian countries have either signed treaty or entered into a Memorandum of Understanding (MoU), however, in many cases growing water stress are posing a challenge to such water arrangements.

India–Pakistan Water Disputes

In 1947, as a result of the partition of British India, India and Pakistan emerged as two sovereign countries. The partition of British India also disturbed the irrigation system developed in Punjab and Bengal during the British's rule. The partition witnessed killing of a large number of people, many women were raped and a number of people were displaced from their homes and moved to the other side of the border.

To maintain the water supply 'Standstill Agreement' was signed in 1947, it expired on 31 March 1948, and on 1 April India stopped water supply to West Pakistan. This matter was resolved after Pandit Jawaharlal Nehru, then prime minister of India, interfered to break the deadlock emerged between the negotiating team from India and Pakistan. Under the agreement, Pakistan had agreed to pay seigniorage charges to India for supply of

---

[28] Suwal, Sahisna Water Crisis in Nepal, *The Water Project*. https://thewaterproject.org/water-crisis/water-in-crisis-nepal. Accessed on 12 June 2018.

[29] Insincere contractor puts Dharahara construction in limbo. *The Rising Nepal*. http://www.therisingnepal.org.np/news/2416. Accessed September 4 2019.

water from East Punjab.[30] The agreement could not fully resolve the disputes. Later, in 1951, David E. Lilienthal, former chairman of Tennessee Valley Authority and the United States Atomic Energy Commission, paid a visit to the Indus region and wrote an article which was published in the popular American magazine *Colliers*. This article influenced Eugene Black, the then President of the World Bank, to offer help of the World Bank to resolve the water disputes between India and Pakistan over the IRS waters.[31]

After years of serious negotiations on 19 September 1960 Indus Waters Treaty (IWT) was signed at Karachi (then capital of Pakistan). The IWT allocates three western rivers to Pakistan—Indus, Jhelum and Chenab—barring some limited uses for India. India got entire waters from the eastern rivers (Ravi, Beas and Sutlej) while Pakistan is allowed to use unused waters from these rivers flowing into its territory. Pakistan cannot make any claim or express rights over the waters from the eastern rivers. Unlike Pakistan, India has limited rights on the western rivers. India is allowed to have about 3.60 Million Acre Feet (MAF) of storage (0.40 MAF on Indus, 1.50 MAF on the Jhelum and 1.70 MAF on the Chenab).[32]

From the day the IWT was signed, it has been facing criticisms in both countries, however, despite three India–Pakistan wars (1965, 1971, and 1999), and a series of political-cum-military tensions between them the IWT remains in force. In recent times, in 2016, Modi made a statement that "water and blood cannot flow simultaneously"[33], after the Pakistan based militant groups killed twenty Indian Army soldiers in Uri sector in the Indian side of Jammu & Kashmir. But soon after, India relented and

---

[30] Haines, D. (2017). *Indus Divided: India, Pakistan and The River Basin Dispute*. New Delhi: Penguin Publications.

[31] Salman, M. A., & Uprety, K. (2003). *Conflicts and Cooperation on south Asia's International Rivers: A Legal Perspective*. Washington, DC: The World Bank.

[32] Ministry of Water Resources & Ganga Rejuvenation. (1960). *Government of India [Indus Waters Treaty]*. Retrieved on September 14, 2018, from http://mowr.gov.in/sites/default/files/INDUS%20WATERS%20TREATY.pdf.

[33] Blood and water can't flow together: PM Narendra Modi gets tough on Indus treaty *The Times of India* 27 September 2016. https://timesofindia.indiatimes.com/india/Blood-and-water-cant-flow-together-PM-Narendra-Modi-gets-tough-on-Indus-treaty/articleshow/54534135.cms. Accessed on 28 September 2016.

participated in the annual Indus Waters Commissioner's meeting between the two countries. In February 2019, once again when India–Pakistan relationships tensed after a militant attack in which 40 soldiers from the Indian paramilitary force, Central Reserve Police Force, lost their lives, the government was under pressure[34] to 'retaliate' against Pakistan from where the Jaish-e-Mohammad (JeM), the organization which took responsibility for the attack, operates. In one of the 'retaliatory' actions the government of India decided to stop the flow of its share of waters from the eastern rivers into Pakistan. This was announced on 21 February 2019 by the Union Minister for Water Resources of India, Nitin Gadkari. This decision will take four to five years to get executed due to lack of water infrastructure. Works on Shahpur Kandi project in Punjab was restarted in 2018 and Ujh multipurpose project in the Indian side of Jammu & Kashmir was finally approved by the Union government of India in January 2019. In reaction, former federal secretary for water resources of Pakistan Khwaja Shumail said that Pakistan has neither concern nor objection to the Indian decision. But the eastern rivers have their significant contribution, though in parts of Pakistan. Sutlej (Beas meets Sutlej in Indian Punjab) enters into Pakistan at Kasur in Punjab while Ravi enters the country near Kartarpur Sahib. In their regions of flow, the waters from the eastern rivers support agricultural and industrial activities, besides fulfilling domestic waters demand of limited population. Also, these rivers are a part of a hydrological unit called Panjnad which is a confluence of five rivers—Jhelum, Chenab, Ravi, and Beas and Sutlej together.

Although Modi government reapproved the Shahpur Kandi project it was planned in 2008 and construction started in 2014. This was mainly to stop the flow of waters from the eastern rivers to meet the water needs of Punjab and Haryana. Even the Ujh project has similar objective to divert waters to satisfy water needs in India.

India–China Water Issues

Growing water demands in mainland China has been always a reason why it is being accused by the lower riparian countries for diversion and stopping the flows to the lower riparian countries. The main river which the two

---

[34] Such pressure was, what Noam Chomsky calls, manufactured by the media houses. Most of the Indian news channels with some notable exceptions call on the government of the day to revoke the IWT and also to declare a war against Pakistan.

DOMESTIC WATER STRESS, TRANSBOUNDARY TENSIONS AND DISPUTES **193**

countries share is the river Brahmaputra or the Yarlung Zangbo River. It originates in Angsi glacier in the southwestern part of Tibet Autonomous Region (TAR) in China. Brahmaputra drains about 580,000 square kilometres of area. Of which 293,000 square kilometres is in TAR, 240,000 square kilometres in India and 47,000 square kilometres is in Bangladesh.[35] On an average, mean annual flow of the river Brahmaputra and its tributaries, from Tibet to India is about 165, 400 cubic kilometre.[36]

Brahmaputra river gets this name only after it enters into plains of Assam and meets its tributaries. In Arunachal Pradesh, Yarlung Zangbo enters through a series of narrow gorges between the mountainous massifs of Gyala Peri and Namcha Barwa. There it is known as Siang and after entering in South Assam it is called Dihang. In south of Pasighat and west of Sadiya town in Assam Dibang and Luhit rivers join Dihang, and they combined are called Brahmaputra.[37] Later the river is also joined by Subansiri, Manas, Sankosh, Teesta, etc. They add about 66% of waters in the river Brahmputra.[38] Therefore, the Chinese structures in the upper stream will have little impact on the flow of river Brahmaputra.

Due to physiography of the region, often landslides occur in the upper stream of the Brahmaputra river. This, often, makes India to allege China for floods in its territory. In 2000 due to landslides in upper stream natural dams were created which dried up the river Siang for a time being.[39] Massive landslide on the Yigong-Tsangpo blocked the river and created a 90-metre deep natural reservoir across 2.5 square kilometres. Two months later this breach of natural dam in Tibet led to severe floods and left over a

---

[35] Ministry of Water Resources, Rivers Development & Ganga Rejuvenation, Government of India. *Annual Report 2016-17-Brahmaputra Board.* http://mowr.gov.in/sites/default/files/AR_BB_2016-17-English.pdf. Accessed April 21, 2018.

[36] Chellaney, B. (2011). *Water: Asia's New Battleground.* Noida: HarperCollins

[37] Bandyopadhyay, J., Ghosh, N., & Mahanta, C. (2016). *IRBM for Brahmaputra Sub-basin: Water Governance, Environmental Security and Human Well-Being.* New Delhi: Observer Research Foundation.

[38] Nilanjan, G. (2017). China Cannot Rob Us from Brahmaputra. *The Hindustan Times.* https://www.thehindubusinessline.com/opinion/china-cannot-rob-us-of-brahmaputra/article9974000.ece. Accessed September 25, 2018.

[39] *India Today.* (2012, March 2). China Diverting Tibet Waters Northwards. https://www.indiatoday.in/world/neighbours/story/china-diverting-tibet-water-northwards-94834-2012-03-02. Accessed March 24, 2017.

hundred people dead or missing in Arunachal Pradesh and Assam.[40] It also swept away large swathes of forest and destroyed over 50 villages in the two mentioned states of India.[41] Like 2000, in 2017 a fear was expressed by India when waters from the river Brahmaputra turned dark due to debris flowing into it because of landslides in upper stream. China refuted such reports that the flow of debris in the Siang river is due to its attempts to build tunnel to divert waters in Xinjiang region[42] In October 2018 also, flash floods occurred in Arunachal Pradesh because of a lake caused by the landslide in the upper stream of the river Brahmaputra. However, the timely information provided by China helped the administration to move the people living near the Siang river in Arunachal Pradesh to safe places.[43]

In the Western sector, on river Sutlej, China has been engaged in the construction of hydro structures. The issue of one the dam on the Chinese side of Sutlej was first raised by the Indian media in March 2006. Reacting to the reports then the Foreign Ministry spokesperson said that 'In order to satisfy the electricity needs of the local population, the Chinese side has built a small-scale hydro-electric station on the Sutlej River at Zada county recently'.[44] However, addressing the fears expressed by the Indian media about the dam, the spokesperson added that 'In the process of development of trans-border rivers, the Chinese side fully considers the impact on its lower reaches'.[45] Also, in the past due to timely information not provided to India flash floods have occurred in Sutlej basin areas. One of them was in the year 2000. In that mishap in the Indian state of Himachal Pradesh about hundred people died, 120 kilometres of strategic highway (in Chini sector) was washed away and ninety-eight bridges destroyed. Four years after that

---

[40] Rediff.com. (2000, July 10). *Chinese Dam Breach Caused Northeast Floods: AFP.* http://www.rediff.com/news/2000/jul/10china.htm.

[41] Mazoomdar, J. (2017, December 22). What's Darkening Brahmaputra: Landslide, Not Chinese Machines. *The Indian Express.* http://indianexpress.com/article/india/whats-darkening-brahmaputra-landslide-not-chinese-machines-4993783/. Accessed April 23, 2018.

[42] Varma, K. J. M. (2017). China to Maintain Communication with India on Artificial Lakes. *Live Mint.* https://www.livemint.com/Politics/Nxmanja68r4b5sqIrj8GhP/China-to-maintain-communication-with-India-on-artificial-lak.html. Accessed February 25, 2018.

[43] Ministry of Foreign Affairs, People's Republic of China. (2018). *Spokesperson Hua Chunying's Regular Press Conference.* https://www.fmprc.gov.cn/mfa_eng/xwfw_665399/s2510_665401/t1606198.shtml (Ministry of Foreign Affairs, PRC).

[44] DNA. (2006). China Defends Dam on Sutlej. https://www.dnaindia.com/world/report-china-defends-dam-on-sutlej-1040220. Accessed August 24, 2018.

[45] Ibid.

mishap, in 2004 once again similar natural processes occurred in upper basin of the river. An artificial lake was formed near river Pareechu in Tibet due to seasonal landslides. This lake was sixty metres deep and had an area of about 230 hectares. To meet the situation a red alert was issued by the Himachal government and armed and paramilitary forces were deployed on a war footing. Nathpa-Jhakri project on river Satluj employing more than 1000 people had to be shut down.[46] Nothing happened in 2004, however, people left their houses in anticipation of repetition of 2000 like disaster.

There is a long debate in China over the use of waters from India flowing rivers. In 2011 China's Vice-Minister of Water Resources, Jiao Yong had publicly stated his government's position against any such project. The vice-minister categorically stated that there are technical difficulties, and it would affect China's relationships with its co-riparian neighbours.[47] Despite the official position of China on the Greater Western Line project, the debate in support of it is still going on. In July 2017, around 20 scholars met outside Urumqi in Northwest China's Xinjiang Uyghur Autonomous Region and discussed the feasibility of diverting waters from Qinghai-Tibet Plateau to Xinjiang's lowland plains. Supporting this project Professor Ren Qunluo said that 'Water from rivers such as Yarlung Zangbo River can help turn the vast deserts and arid lands into oasis and farmlands, alleviate population pressure in the east, as well as reduce flood risks in the countries through which the river travels downstream'.[48]

Cooperative arrangements between India and China over water issues were first signed in 2002 and 2005, respectively when China agreed to provide hydrological information of major transboundary rivers to India in flood season. In 2010, an MoU was signed under which China agreed to share water-related information of river Langquen Zangbo/Sutlej with India. In 2015, during the visit by the Indian Vice-President, Hamid Ansari

---

[46] Yogendr, K. (2004, August 13). A flood of Rumours. *The Hindu.* http://www.thehindu.com/2004/08/13/stories/2004081307341100.htm. Accessed July 12, 2016.

[47] Wirsing, R. G., Stoll, D. C., & Jasparro, C. (2013). *Transnational Conflict Over Water Resources in Himalayan Asia.* New York: Palgrave Macmillan.

[48] *Global Times.* (2017). Scholars Mull Project to Divert Water from Tibet to Arid Xinjiang. http://www.globaltimes.cn/content/1060047.shtml. Accessed September 12, 2018.

to Bejing, the MoU was renewed for further five years. An implementation plan to the MoU was signed in 2016 during the 10th Expert Level Mechanism meeting at New Delhi.[49]

In 2008 India and China signed cooperation on the provision of Hydrological Information of the Brahmaputra/Yarlung Zangbo River in Flood Season by China to India.[50]

However, this hydrological data information sharing MoU between India and China could not sustain the pressure of India–China military stand-off in Doklam in 2017. During the stand-off this information sharing process was interrupted[51] It was resumed in March 2018.

Significance of Transboundary Waters in India–Bangladesh Relations

India and Bangladesh share 54 rivers between them. Disputes over the proposal to build Farakka barrage erupted when Bangladesh was still a part of Pakistan. The main objective of the barrage was to ensure that the river Hooghly receives up to 40,000 cubic feet per second (cusecs) of water, no matter how low the flow of water in Ganges remains.[52]

After the liberation of Bangladesh in 1971, in 1972 India and Bangladesh signed friendship treaty under which the two sides also agreed to find an amicable solution to their water-related issues.

Bangladesh contested that due to existence of Farakka Barrage, it had to face multiple problems. In the dry season which spans from January to May, drought or drought like situation prevails almost every year. One of the reasons for such situation in Bangladesh, as believed by many in the country, is diversion of waters through feeder canals from Farakka barrage to Hooghly. On the contrary, during the monsoon which lasts from June to September, Bangladesh faces severe flood, when the melting snow of

---

[49] Ministry of Water Resources, Rivers Development & Ganga Rejuvenation, Government of India. *INDIA-CHINA COOPERATION.* http://mowr.gov.in/international-cooperation/bilateral-cooperation-with-neighbouring-countries/india-china-cooperation. Accessed April 12, 2018)

[50] On condition of anonymity an officer from Central Water Commission told the author that many times the data India gets does not match with the real flow in the river. Also, it is not known how much water is available in the upper region of the transboundary rivers.

[51] Gupta, J. (2018). *China, India, and Water Across the Border.* https://www.thethirdpole.net/en/2018/05/01/china-india-and-water-across-the-border/. Accessed October 12, 2018.

[52] Salman, M. A., & Uprety, K. (2002). *Conflicts and Cooperation on south Asia's International Rivers: A Legal Perspective.* Washington, DC: The World Bank.

DOMESTIC WATER STRESS, TRANSBOUNDARY TENSIONS AND DISPUTES 197

the Himalayas and the heavy rain in the region reach Bangladesh through the three mighty rivers and the smaller rivers, on their way to drain at the Bay of Bengal. About 2.6 to 3 million hectares(in Bangladesh) were flooded annually. In an abnormal year, when there is a synchronization of very heavy rainfall with peak discharges simultaneously in the Ganga and Brahmaputra, this figure may reach 6.5 million hectares or some 45% of the total area as happened in 1955 and 1974.[53]

To address their water issues, in 1975 partial accord was signed between India and Bangladesh. The accord expired on May 31, 1975. As it was not replaced by a new accord or agreement, India began withdrawing to the full capacity of the feeder canal of 40,000 cusecs.[54]

Under General Ziaur Rehman, Bangladesh used international platforms to raise India's water hegemony. It raised the matter in the seventh Islamic Foreign Ministers' Conference, which was held in Istanbul, Turkey,[55] and also at the United Nations. At UN Consensus Statement was adopted on 26 November 1976.[56] Subsequently, in 1977 the two countries signed an agreement to share waters from river Ganga. This agreement was for five years but, practically, it ended on 31 May 1982.[57] After a series of MoU to renew the water sharing arrangements, finally in 1996 India and Bangladesh signed Ganga Water sharing Treaty.

The 1996 Ganga water sharing treaty has not found favour from many in West Bengal. In recent years, In May 2017, in a letter to the Indian Prime Minister Mamata Banerjee, the Chief Minister of Indian state of West Bengal said, 'West Bengal government's experience with the 1996 India–Bangladesh Ganga water sharing treaty was not a happy one'.[58] In the letter 'She pointed out the "adverse" impacts on the availability of water in her state and land erosion in Malda, Murshidabad and Nadia districts. The lack of water in the Ganga during the lean season occasionally causes

---

[53] Bhasin, A. S. (Ed.). (1996). *India-Bangladesh Relations,1971–1994, Documents* (p. 88). Delhi: Siba Exim Pvt. Ltd.

[54] Ibid.

[55] Khan, Z. A. (1976). *Basic Documents On Farakka Conspiracy* (p. 152). Dacca: Khoshroz Kitab Mahal.

[56] Bhasin, A. S. (Ed.). (1996). *India-Bangladesh Relations,1971–1994, Documents*. Delhi: Siba Exim Pvt. Ltd.

[57] Ibid.

[58] *The Daily Star*. (2017, May 26). Mamata at It Again. http://www.thedailystar.net/frontpage/mamata-hits-back-1410946. Accessed May 26, 2017.

shutting down of the National Thermal Power Plant in Farakka. …The promise of making water available has not been fulfilled'.[59]

More than Ganga waters, now India and Bangladesh have tensions over Teesta river waters. In 2011, India and Bangladesh signed an interim agreement to share Teesta's waters between them. Under it, India would receive 42.5% and Bangladesh 37.5% of water during the lean season (December to May). This is not acceptable to the Mamata Banerjee. One of the reasons for West Bengal's reluctance to agree on the present deal is: Water from Teesta is important for the irrigation in the five districts of the north Bengal—Coochbehar, Jalpaiguri, South and North Dinajpur, Darjeeling—which constitute some of the poorest blocks in the state. With viable irrigation system these areas have the capability to produce three crops in a season.[60] A related reason is decline in the water table due to the proliferation of tea plantation industry in Darjeeling and Jalpaiguri.[61] India and Bangladesh also have issues on rivers such as Feni, Manu, Muhuri, Khowai, Gumti, Dharla and Dudhkumar, which they are negotiating.

India and Bangladesh are also looking at all other aspects of the issues on all 54 rivers they share. During his visit to Dhaka from 19 to 21 August, 2019 India's External Affairs Minister said that 'We are [India and Bangladesh] ready to start from anywhere……We look forward to making progress in finding a mutually acceptable formula to share water from 54 shared rivers'.[62]

India's Water-Related Tensions with Bhutan and Nepal

More than water sharing, India's issues with Nepal and Bhutan are mainly because of hydropower projects and other irrigation projects developed by India in the respective countries. Drangme Chhu is the largest river system of Bhutan. It rises in the Indian state of Arunachal Pradesh. There are about

[59] Ibid.

[60] Das, M. (2015, June 13). Teesta Accord: West Bengal CM Mamata Banerjee May Be Eyeing Bigger Compensation. *The Economic Times*. http://economictimes.indiatimes.com/news/politics-and-nation/teesta-accord-west-bengal-cm-mamata-.

[61] Ibid.

[62] "Let's find a mutually acceptable formula." *The Daily Star*, 2019, 21 August. https://www.thedailystar.net/frontpage/news/lets-find-mutually-acceptable-formula-1788139. Accessed on 22 August 2019.

DOMESTIC WATER STRESS, TRANSBOUNDARY TENSIONS AND DISPUTES **199**

56 rivers which flow down from Bhutan to the Indian state of Assam to meet the Brahmaputra River.[63]

As a water-rich country, Bhutan has a capability to produce 30,000 Mega Watts of hydroelectricity every year, of which, only about 1616 megawatts is being currently generated.[64] To develop its hydro electricity generation capacity, India has been assisting Bhutan since 1961. In 2006, the two countries signed an agreement where India agreed to develop a capacity of 5000 MW by 2020. In 2009, an additional protocol to it was signed under which the projected capacity was increased to 10,000 MW by 2020. However, there remain concerns among the people from Bhutan due to growing hydro debt and other related impacts of the projects in the region.

In 2016, a New Delhi-based NGO Vasudha Foundation[65] came out with a report highlighting the environmental, economic, political and social concerns of the Indian hydropower projects in Bhutan. The report further substantiated the hydropower debt Bhutan is facing due to Indian hydropower projects. However, on the findings, then Bhutanese Prime Minister Tshering Tobgay said that 'If they are concerned about our environment, I am also concerned. They should have met concerned people from the government before publishing'.[66] He told thethirdpole.net that 'I agree hydropower projects do damage the environment. But the hydropower projects in Bhutan, which are run-of the river schemes, do the least damage'.[67] He added, 'When we have hydropower, we do not have to use thermal power, or power generated from nuclear sources or coal. That is good for the environment. In that context hydropower is good for environment'.[68]

---

[63] Yashwant, S. (2018, August 27). *Villagers in Bhutan and India Come Together to Share River.* Thethirdpole.net. https://www.thethirdpole.net/en/2018/08/27/villagers-in-bhutan-and-india-come-together-to-share-river/.

[64] Asian Development Bank, *Water Securing Bhutan's Future*, 2016. https://www.adb.org/sites/default/files/publication/190540/water-bhutan-future.pdf. Accessed on 2 April 2018.

[65] Vasudha Foundation. *A Study of the India-Bhutan Energy Cooperation Agree ments and the Implementation of Hydropower Projects in Bhutan.* Vasudha Foundation, January 2016. http://www.vasudha-foundation.org/wp-content/uploads/Final-Bhutan-Report_30th-Mar-2016.pdf. Accessed March 28, 2018.

[66] Walker, B. (2016, October 4). *Bhutan's PM Defends Hydropower Dams Against Blistering Report.* Thethirdpole.net. https://www.thethirdpole.net/en/2016/10/04/bhutans-pm-defends-hydropower-dams-against-blistering-report/.

[67] Ibid.

[68] Ibid.

The Bhutanese complain that India buys cheap electricity from the hydroelectric projects in Bhutan. For example, in 2017, the tariff rate on the import of hydroelectricity from the Tala hydroelectric project by India was 1.80 Bhutanese Ngultrum (BTN) per unit. This was much below the domestic market price in India which was around ₹7–₹8.[69] Then the 2016 Cross Border Trade of Electricity (CBTE) guidelines is seen unfavourable to Bhutan. In clause 5.2.1 it states[70]:

a. Import of electricity by Indian entities from Generation projects located outside India and owned or funded by [the] Government of India or by Indian Public Sector Units or by private companies with 51% or more Indian entity (entities) ownership.
b. Import of electricity by Indian entities from projects having 100% equity by Indian entity and/or the Government/Government-owned or controlled company(ies) of [a] neighbouring country.
c. Import of electricity by Indian entities from licenced traders of neighbouring countries having more than 51% Indian entity(ies) ownership, from the sources as indicated in paras. 5.2.1(a) and 5.2.1(b) above.
d. Export of electricity by distribution licensees/Public Sector Undertakings (PSUs), if surplus capacity is available and certified by the concerned distribution licensee or the PSU as the case may be.

Third, with the implementation of the Goods and Services Tax in India in 2017, its exports to Bhutan are cheaper than the imports from Bhutan. This will have an impact on Bhutan's trade deficit with India.[71] Due to

[69] Tenzing, L. (2018, July 26). More Than Doklam Issue, Bhutan Worries About Hydropower Deficits, *The Indian Express*. http://indianexpress.com/article/opinion/more-than-the-doklam-issue-bhutanworried-about-hydropower-deficits-4768598/. Accessed July 27, 2017.

[70] Ministry of Power, Government of India. *Guidelines on Cross Border Trade of Electricity*. https://powermin.nic.in/sites/default/files/webform/notices/Guidelines_for_Cross_Boarder_Trade.pdf. Accessed March 27, 2018.

[71] Tenzing, L. (2018, July 26). More Than Doklam Issue, Bhutan Worries About Hydropower Deficits. *The Indian Express*. http://indianexpress.com/article/opinion/more-than-the-doklam-issue-bhutanworried-about-hydropower-deficits-4768598/. Accessed July 27, 2017.

differences over the guidelines, few projects such as Kholongchhu were stopped. In 2018, amendments were made in the guidelines and now the 51% rule has been removed. Afterwards, work on the installed projects has started.

Like, Bhutan, India–Nepal water-related tensions are also mainly because of Indian hydropower projects in Nepal. India and Nepal share about 6000 water bodies. In 1996 after the Mahakali treaty was signed, a section of Nepalese raised various objections over it. Under the treaty, India has to develop hydropower projects in the country.

One of the early water-related agreement India and Nepal signed was in 1954 on river Kosi. Soon after its signing, the treaty was criticized in Nepal.[72] Critics said that the projects on the river gives extraterritorial rights to India for an indefinite period without providing Nepal with adequate compensation. Also, it was stated that the scheme was actually designed for the furtherance of India's own interests without paying proper attention to the well-being of the Nepalese people.[73] Under pressure, in 1966 this treaty was amended. In the amended treaty it was clarified by Nepal that 'that the Government of India will be reasonably compensated in case the Project properties are taken over by His Majesty's Government at the end of the lease period. The compensation will cover the cost borne to date and such other cost as may be incurred in future by the Government of India with the agreement of His Majesty's Government. In that case the depreciation in the value of the Project materials would, of course, be taken into account'.[74]

In December 1959, Gandaki river Agreement was signed between India and Nepal.[75] Gandaki project agreement called on for construction of a barrage, canal head regulators and other appurtenant works of about 1000 feet below the Tribeni canal head regulator.[76] This too was criticized and

---

[72] Salman, M. A., & Uprety, K. (2002). *Conflicts and Cooperation on South Asia's International Rivers: A Legal Perspective.* Washington, DC: The World Bank.

[73] Ibid.

[74] Ministry of Economic Affairs, Government of Nepal. (1966). https://www.internationalrivers.org/sites/default/files/attached-files/treaties_between_nepal-india.pdf.

[75] Salman, M. A., & Uprety, K. (2002). *Conflicts and Cooperation on South Asia's International Rivers: A Legal Perspective* (p. 83). Washington, DC: The World Bank.

[76] Ibid., 84.

was amended in 1964. The amended treaty deleted clause 10 and clause 9 was modified to read as: 'His Majesty's Government will continue to have the right to withdraw for irrigation or any other purpose from the river or its tributaries in Nepal such supplies of water as may be required by them from time to time in the Valley'.[77] In recent years, India is engaged in the construction of a number of projects such as Arun 3, upper Karnali, etc. Some of those such as the one on rivers Kosi and Karnali have faced opposition due to which they are being delayed.

Water Issues between Afghanistan and Pakistan

Among all rivers shared between Pakistan and Afghanistan, Kabul river is the most important. It rises in Hindu Kush and drain into the river Indus near Attock in Pakistan. This river fulfils around 26% of the Afghanistan's freshwaters supply, and around 25 million of people are depended on river Kabul and its tributaries for their livelihood.[78] In Pakistan, Kabul river and its tributaries are indispensable to quench the thirst of more than two million people from Peshawar, and for irrigation purposes in Peshawar valley and North Waziristan. Its waters are also needed to run 250 MW hydropower dam in Warsak.[79]

As the two countries are depended on the waters from the Kabul river system, they want to augment more quantity of waters to satisfy their demands. In 2018, India agreed to assist Afghanistan in building $236 million Shahtoot Dam on the Kabul River. This is expected to hold around 146 cubic metres of waters for drinking for about 2 million residents of Kabul. It is also expected to irrigate around 4000 hectares of land. This

---

[77] Ministry of Economic Affairs, Government of Nepal. (1966). *Gandak Water Treaty.* https://www.internationalrivers.org/sites/default/files/attached-files/treaties_between_nepal-india.pdf.

[78] Majidyar, W. (2018, December 15). Afghanistan and Pakistan's Looming Water Conflict. *The Diplomat.* https://thediplomat.com/2018/12/afghanistan-and-pakistans-looming-water-conflict/. Accessed April 5, 2019.

[79] Kakakhel, S. (2017, March 2). *Afghanistan-Pakistan Treaty on the Kabul River Basin?* Thethirdpole.net. https://www.thethirdpole.net/en/2017/03/02/afghanistan-pakistan-treaty-on-the-kabul-river-basin/. Accessed April 5, 2019.

is a part of 12 dams which Afghanistan is planning to build in Kabul river basin with Indian assistance.[80]

Pakistan has a problem with Shahtoot dam project. It is estimated that around 17 million acre feet of waters is received by Pakistan every year from Kabul river.[81] After the construction of all 12 dams there will be, around 16–17% drop in water flow in Pakistan.[82]

At present, there is no treaty between Pakistan and Afghanistan. The water flows between them are guided through the principles of international riparian laws, conventions and customs. In 2006 the United States Agency for International Development and the World Bank supported the idea of having water sharing treaty between Pakistan and Afghanistan.[83] Yet the two countries did not. In the absence of treaty the water tensions between them are likely to escalate in future.

CONCLUSION

As discussed in this chapter, rise in water demands in the Himalayan South Asia increase the burden on transboundary rivers. Most of the time the upper riparian raises the issue of having 'sovereign ownership' over those waters while the lower riparian states talk about right of historical users and give evidence of international laws which secures the rights of lower riparian states. The condition is serious where the countries have not signed water sharing treaties. However, the countries which have water sharing treaties between them also have tensions and, mainly, due to political reasons and the rising demands for water.

[80] Majidyar, W. (2018, December 15). Afghanistan and Pakistan's Looming Water Conflict. *The Diplomat.* https://thediplomat.com/2018/12/afghanistan-and-pakistans-looming-water-conflict/. Accessed April 5, 2019.

[81] *Sharing Water Resources with Afghanistan. Dawn,* 13 November 2011. https://www.dawn.com/news/673055. Accessed January 12, 2019.

[82] Hessami, E. (2018, November 13). Afghanistan's Rivers Could Be India's Next Weapon Against Pakistan. *Foreign Policy.* https://foreignpolicy.com/2018/11/13/afghanistans-rivers-could-be-indias-next-weapon-against-pakistan-water-wars-hydropower-hydrodiplomacy/. Accessed April 5, 2019.

[83] Awan, H. M. A. (2018, July 9). A Pak-Afghan Water Treaty? *The News.* https://www.thenews.com.pk/print/339335-a-pak-afghan-water-treaty.

In the coming years, as there will be a rise in demands for water, disputes over transboundary waters would likely aggravate. To manage such situations, alternatives have to be found out in terms of managing the rising demand by giving stress on innovating means to get more crops per drop, etc. There is also a need to have cooperative water arrangements in the region to manage the available water resources.

For a large number of Himalayan South Asian rivers, Tibet in China is the source of origin. As China itself faces water scarcity, it is being accused for disturbing the water flows by diverting or choking the water flows in the upper stream. China's engagement in the hydropower projects in countries such as Pakistan and Nepal have its repercussions on the country's relationships with India. For example, China's investment under China Pakistan Economic Corridor in the hydropower sector is being opposed by India which finds that they are 'infringement' to India's sovereignty which considers Pakistan administered Kashmir as occupied territory. In Nepal too China is investing about $ 2.5 billion USD to build hydroelectric project on Gandaki river. Given the relationships with China, India is also cautious about this project, though it has not made any strong official statement. China's geographical location as upper riparian to the major Himalayan South Asian river system and growing water stress within China may influence its transboundary water behaviour in future. And this is a matter of concern for the lower riparian States in the Himalayan South Asia.

BIBLIOGRAPHY

Ahmad, Q. K., Biswas, A. K., Rangachari, R., & Sainju, M. M. (Eds.). (2001). *Ganges-Brahmaputra-Meghna Region: A Framework for Sustainable Development.* Dhaka: University Press Limited.

Azim Uddin, A. F. M., & Baten, M. A. (2011, October). *Water Supply of Dhaka City: Murky Future the Issue of Access and Inquality.* Dhaka: Unnayan Onneshan the Innovators, Centre for Research and Action on Development.

Basu, M., & Shaw, R. (2013). Water Policy, Climate Change and Adaptation in South Asia. *International Journal of Environmental Studies, 70*(2), 175–191.

Bhutan. Retrieved from http://www.fao.org/nr/water/aquastat/countries_regions/BTN/.

Chowdhury, N. T. (2010). Water Management in Bangladesh: An Analytical Review. *Water Policy, 12,* 32–51.

Goldar, A. (2019, June 8). *Water Policy and Development Perspective in Bangladesh: A Critical Review in Bangladesh.* (Unpublished Mimeo). Retrieved from https://www.academia.edu/27023531/Water_Policy_and_Development_Perspective_in_Bangladesh_A_Critical_Review_in_Bangladesh.

Hasan, S., & Mulamoottil, G. (1994). Natural Resource Management in Bangladesh. *Ambio, 23*(2), 141–145.

Hossain, M. A. A., Weng, T. K., & Mokhtar, M. B. (2012). *Water Institutional Change in Bangladesh: Experiences and Future Needs.* New York: Nova Science Publishers.

Huq, S., Karim, Z., Asaduzzaman, M., & Mahtab, F. (Eds.). (1999). *Vulnerability and Adaptation to Climate Change for Bangladesh.* Dordrecht: Kluwer Academic Publishers.

© The Editor(s) (if applicable) and The Author(s), under exclusive license to Springer Nature Singapore Pte Ltd. 2020
A. Ranjan (ed.), *Water Issues in Himalayan South Asia,*
https://doi.org/10.1007/978-981-32-9614-5

206 BIBLIOGRAPHY

Khuda, Z. R. M. M. (2001). *Environmental Degradation Challenges of the 21st Century*. Dhaka, Bangladesh: Environmental Survey and Research Unit.

Kuensel. (2017, July 3). *Water Scarcity in Water Rich Bhutan*. Retrieved from http:// www.kuenselonline.com/water-scarcity-in-water-rich-bhutan/.

Kuensel. (2018, June 18). *Water Shortage in Water Abundant Bhutan*. Retrieved from http://www.kuenselonline.com/water-shortage-in-water-abundant-bhutan/.

Nabi, R. (2018, October 2). Bhutan's Happiness Saga. *South Asia Journal*. Retrieved from http://southasiajournal.net/bhutans-happiness-saga/.

National Statistics Bureau. (2019). Retrieved from http://www.nsb.gov.bt/main/main.php.

Rahman, M. M. (2005, December 9–10). *Bangladesh—From a Country of Flood to a Country of Water Scarcity—Sustainable Perspective for Solution*. Seminar on Environment and Development, Hamburg, Germany, Entwicklungsforum Bangladesh e.V.

South Asia Journal. (2011, August 31). *Save the Rivers to Save Yourselves*. Retrieved from http://southasiajournal.net/nurul-huq-save-the-rivers-to-save-yourselves/.

South Asia Network on Dams, Rivers and People. Retrieved from https://sandrp.in/tag/min-istry-of-water-resources/.

Statistical Yearbook. (2014). *National Statistical Bureau*. Bhutan. Retrieved from http://www.nsb.gov.bt/publication/files/yearbook2014.pdf.

WARPO (Water Resources Planning Organisation). (1999). *National Water Management Plan*. Dhaka, Bangladesh: WARPO.

Water Resources Management Plan. (2003). *Ministry of Economic Affairs*. Bhutan.

Water Scarcity in Bangladesh: Transboundary Rivers, Conflict and Cooperation (PRIO Report 1). (2013).

World Bank. (2000). *Bangladesh: Climate Change and Sustainable Development* (Bangladesh Report No. 21104-BD). Rural Development Unit, South Asia Region, Document of the World Bank.

INDEX

A

Agriculture, 2, 3, 10, 11, 17, 22, 31, 34, 35, 40, 45, 49, 52, 64, 66, 67, 91, 116, 145, 157, 165, 166, 168, 169, 171, 173, 180, 181, 188

Arsenic, 38, 40, 43, 176, 187

Arun, 128, 132, 135–139, 148–150

B

Basin, 2, 6, 9, 15, 18, 27, 30, 39, 41, 46, 54, 66, 96, 97, 157, 158, 160, 163–165, 170, 176, 178, 194, 195

Benefit sharing, 128, 129, 139, 142, 143, 148–152

C

Chhukha, 54

Climate change, 3–5, 14–16, 24, 29, 30, 41, 42, 48, 56, 57, 63, 70, 74, 75, 78, 95–97, 100, 101, 111, 155, 159, 162, 164, 168, 183, 189

Collaborative governance, 96

Commodity, 37, 104, 106, 108, 111, 120, 123

Commons, 104, 108, 120, 123

Conflict resolution, 147, 152

Consumption, 2, 4, 7, 10–12, 14, 22, 40, 88–90, 126, 142, 147, 157, 176, 184, 186, 187

D

Dams, 18, 37, 57, 66, 67, 69, 72, 107, 111, 113, 116–118, 121–123, 139, 140, 154, 162, 164, 165, 170, 178–180, 186, 193, 194, 202, 203

Development discourse, 106, 108, 123

Dispute, 115, 119, 123, 127–130, 132, 135, 138, 142, 143, 145–152, 164

Drinking, 6, 15, 17, 22, 23, 32, 38, 42, 43, 48, 57, 64, 67, 68, 70, 71, 74, 107, 111, 113, 120, 129, 176, 177, 185, 202

© The Editor(s) (if applicable) and The Author(s), under exclusive license to Springer Nature Singapore Pte Ltd. 2020
A. Ranjan (ed.), *Water Issues in Himalayan South Asia*,
https://doi.org/10.1007/978-981-32-9614-5

208 INDEX

Dry, 8, 15, 29, 31–33, 36, 38, 40, 41,
44, 55, 69, 70, 126, 154, 177,
186, 196

E
Entitlement, 108, 120, 123
Extraction, 14, 15, 17, 26, 56, 74, 90,
170–172, 174

F
Floodplain, 32, 38

G
Ganga, 30–34, 38, 40–42, 48, 49, 197,
198
Glacier, 41, 53, 55, 56, 58, 64, 74,
157, 193
Governance, 3, 8, 10, 21, 22, 24, 26,
27, 31, 32, 53, 63, 74, 77–79,
85–87, 89–93, 100–102, 114
Groundwater, 6–8, 14–17, 21–23, 26,
30, 31, 35, 39, 40, 43, 45, 49, 53,
56, 106, 110–112, 116, 119, 123,
158, 160, 162, 169–172, 175,
176, 186

H
Himalayan, 32, 41, 51, 52, 55, 57,
177, 183, 184, 190, 203, 204
Hydroelectricity, 66
Hydropower, 11, 37, 56–58, 61–67,
74, 125–129, 131–135, 137–139,
143, 147–152, 157, 162, 164,
198, 199, 201, 202, 204

I
Indus, 1, 9, 16, 22, 41, 156–160, 162,
163, 165, 170, 175–179, 191,
192, 202

Indus rivers system, 1
Infrastructure, 14, 16–19, 21, 22, 34,
38, 42, 43, 59, 66, 100, 107, 137,
138, 145, 149, 161, 163, 168,
175, 192
Intersectoral, 11
Irrigation, 3, 4, 6, 9, 10, 12–14, 17,
18, 21, 24–26, 31, 33, 36, 38–40,
43–46, 48, 52, 55, 64, 69, 70, 73,
87, 95, 109, 116, 117, 122, 123,
129, 131, 134, 150, 158, 161,
165–168, 170, 173, 185, 190,
191, 198, 202

K
Kabul, 1, 2, 7–10, 12, 14, 15, 18, 19,
22–24, 175, 202, 203
Karnali, 41, 128, 132

N
Narmada, 113, 114, 116, 117, 121,
122

P
Pearl River Delta (PRD), 83, 84, 86,
88, 95, 97, 98, 101, 102
Pollution, 3, 15, 22, 36, 39, 40, 43,
46, 109, 111–113, 121, 123, 186,
188, 189
Population, 3, 6–9, 11, 14–16, 20–23,
26, 29, 31, 32, 34, 37, 40, 41, 43,
52, 58, 62, 63, 66, 67, 69, 81–83,
110, 111, 126, 151, 155, 156,
162, 166, 167, 170, 177, 178,
183, 184, 186, 188, 189, 192,
194, 195
Public policy, 44

R

Rain, 6, 31, 32, 35, 39, 41, 42, 46, 73, 96, 158, 168, 197

River basin, 1, 2, 9–11, 15, 16, 18, 24, 25, 27, 32, 41, 48, 50, 53, 55, 72, 87, 170, 177, 188, 203

S

Sanitation, 22, 23, 44, 69, 111, 173–175, 178, 185

T

Tamakoshi, 128, 133, 143, 144, 146, 147, 149, 151

Transboundary, 1, 9, 11, 18, 27, 28, 44, 46, 47, 60, 72, 75, 77–80, 85–88, 90–93, 98, 100–102, 123, 163, 177, 190, 195, 203

Treaty, 15, 28, 38, 140, 141, 161–164, 190, 196, 197, 201–203

W

Water management, 9, 24–26, 30, 38, 39, 44, 45, 50, 62, 70, 79, 84, 85, 87, 93, 95, 98, 100, 102, 107, 116, 154, 157, 158

Water policy, 45, 64, 67, 71, 74, 102, 110, 119, 159, 166, 188

Water scarcity, 3, 12, 13, 19, 74, 80, 96, 110, 111, 155–157, 174, 183, 185, 186, 204

World Bank (WB), 19, 48, 105, 119, 121, 135–137, 161–164, 173, 174, 191, 203

Printed in the United States
By Bookmasters